Trim Carpentry

Trim Carpentry

Philip Moon

THOMSON

DELMAR LEARNING

Australia   Brazil   Canada   Mexico   Singapore   Spain   United Kingdom   United States

Trim Carpentry
Philip Moon

Vice President, Technology and Trades ABU:
David Garza

Director of Learning Solutions:
Sandy Clark

Senior Acquisitions Editor:
James DeVoe

Product Manager:
Jennifer A. Thompson

Marketing Director:
Deborah S. Yarnell

Marketing Coordinator:
Mark Pierro

Director of Production:
Patty Stephan

Senior Production Manager:
Larry Main

Content Project Manager:
Jennifer Hanley

Art & Design Coordinator:
Nicole Stagg

Technology Project Manager:
Linda Verde

Editorial Assistant:
Tom Best

Library of Congress
Cataloging-in-Publication Data

Moon, Philip.
    Trim carpentry / Philip Moon.--1st ed.
        p. cm.
    Includes bibliographical references and index.
    1. Trim carpentry.   I. Title.
    TH5695.M66 2008
    694'.6--dc22

                                    2006015221

ISBN: 1-4180-2864-9

NOTICE TO THE READER

Contents

Procedures

Preface

INTENDED USE OF THIS BOOK

Designed for carpentry and cabinet-making courses, as well as for anyone desiring to learn this highly regarded craft, *Trim Carpentry* describes in detail almost every aspect of working with interior trim. Based on years of accumulated knowledge and experience, this book highlights the techniques and methods used in the preparation and installation of detail work such as trimming windows and doors, finishing staircases, and installing cabinets, moldings, columns, and hardwood flooring.

HOW TO USE THIS BOOK

Trim carpentry takes place near the end of a construction project. It is one of the final stages. The project, or the building being constructed, may be a new custom home, office building, condominium, hotel, vacation resort, apartment complex, or any other structure requiring the application of interior molding or trim work. Trim carpentry (also referred to as *finish carpentry*) is a trade that demands attention to detail and requires a high level of *skill* (i.e., ability gained by practice or knowledge).

In order to acquire a high level of skill, this book gives readers the opportunity to efficiently learn the techniques and methods that professional trim carpenters use on a daily basis. Reader-friendly chapters provide need-to-know information required for the successful completion of the following tasks of the trim carpenter:

- Install and trim interior doors
- Cut and install base molding
- Case and trim windows
- Install wainscot and wall frames
- Finish closets (shelves, shoeboxes, and hanging rods)
- Cut and install crown molding
- Create ceiling beams
- Finish staircases and balustrades
- Make and install columns
- Install cabinets
- Install hardwood flooring
- Complete countertops

The duties of a trim carpenter, from a contractual point of view, may vary slightly from region to region. For example, a trim carpenter in one part of the country may be expected to install kitchen cabinets or hardwood flooring, while in another part of the country those things may be assigned to a different contractor. For the most part, however, trim carpenters are expected to perform these tasks with proficiency.

FEATURES OF THIS BOOK

- **Logical organization:** Each chapter thoroughly covers one aspect of trim carpentry so that the reader can study and learn about one task before moving on to another subject. The chapters step you through the natural progression of a typical construction job, making it easy to reference a specific method or technique.

- **Extensive photos and illustrations** assist readers in visualizing important tasks and the proper techniques for successfully completing each job.

- **Safety issues** are emphasized in each chapter as important reminders of the types of safety precautions that should be taken during a particular procedure.

- **A Procedures section** in each chapter takes the reader through a particular task in a step-by-step format, making it easy to learn and understand that task.

- **Key Terms** are highlighted in bold type in each chapter, providing readers with the terminology to effectively communicate and understand the tasks at hand on the job.

  Important Note: Many of the terms used in trim carpentry are the same nationwide. However, some terms will vary from one region to another. In other words, several different terms may be used to describe a single material or method. It is important that readers take time to learn the terminology that is used in the region in which they plan to work to ensure clear communication of thoughts and ideas.

- **Review questions** appear at the end of each chapter, allowing readers to evaluate their knowledge and comprehension of the tasks presented in the chapter.

SUPPLEMENT TO THIS BOOK

For those intending to use this book in the classroom environment, an accompanying **Instructor's Guide on CD-ROM** is available. Designed to assist instructors in preparing for classroom presentations, the CD-ROM contains the following features:

- **Answers to Review Questions** provide a quick reference for student evaluation.

- **PowerPoint** outlines of each chapter, as well as art from the book, work to enhance classroom presentations.

- **Chapter Quizzes** provide the instructor with the necessary tools for evaluating student knowledge of the content presented in each chapter.

- **Exercises** highlight additional classroom activities to encourage application of important trim carpentry methods as well as to review important concepts.

(Order #: 1-4180-3239-5)

ACKNOWLEDGEMENTS

I'm grateful to have been given the chance to share what I have learned in the field of trim carpentry. I hope that many others can benefit from the textbook and the subjects covered. Trim carpentry is a very rewarding trade, one that allows you to stop at the end of the day and actually see what you have accomplished with your own hands. I hope those who study this textbook will get a chance to experience that type of satisfaction in their own careers.

Uhai Publishing made the idea of a trim carpentry textbook into a reality. I'd like to thank Mark Huth for his guidance, counseling, and patience. His knowledge and understanding helped make the book what it is, as did the wise comments from the experts who reviewed this book and made suggestions for improvements:

Ramona Cook
Ivy Tech Community College,
Williamsburg, Indiana

David McCosby
New Castle School of Trades,
Pulaski, Pennsylvania

Rick Miles
Williamsport, Pennsylvania

Patrick Molahan
Madison Area Technical College,
Madison, Wisconsin

Michael Nauth
Algonquin College, Ontario, Canada

Sean Quinn
Lansing Community College,
Okemos, Michigan

Terry Schaefer
Western Wisconsin Technical College,
La Crosse, Wisconsin

Lester Stackpole
Eastern Maine Technical and
Community College, Bucksport, Maine

Loran Stara
Southeast Community College,
Milford, Nebraska

Katherine Swan
Tacoma, Washington

Richard V. Telech
Henderson, Nevada

Linda Ireland, who edited the textbook, did a great job of shaping and organizing the manuscript. Margaret Berson completed the final stages of the editing. Thanks to my friend Steve Herman for introducing me to Mark Huth and Uhai Publishing. Thanks also to the Delmar Learning team who took the manuscript and shaped it into the final book— Alison Weintraub, Jennifer Thompson, and Jennifer Hanley. Without these people, the book would not have come into existence.

A special thanks goes to all the carpenters I've had a chance to learn from over the years, even those in Florida: There are too many to name individually, so I thank all of them collectively. And thanks to my wife, Michelle, for enduring my emotional ups and downs for the three years that it took to complete this project.

Thank you,
Phil Moon

ABOUT THE AUTHOR

Phil Moon has spent many years mastering the art of trim carpentry. He currently owns and operates a successful business specializing in custom interior trim and interior trim design, and has worked with master trim carpenters from all over the world. His work can often be found in Florida travel magazines for work he has completed for some of the finest hotels, resorts, and condominiums in and around the Orlando area. Mr. Moon hopes this book teaches and benefits apprentices, contractors, builders and homeowners who are interested in learning more about interior trim carpentry.

Safety First, Last, and Always

What follows is a list of important safety rules for trim carpentry. Special safety concerns are also listed throughout the book. We highly recommend that you read and follow these rules to avoid injury while on the job. While this list is comprehensive, it cannot cover every possible situation, and you should always refer to instructions from the specific manufacturer(s) when operating tools and machinery. The author and Thomson Delmar Learning are committed to keeping you safe.

- Check safety features on a regular basis to make sure they are in proper working order.

- Do not overwork a machine by cutting material at a faster rate than it can handle.

- Keep sharp blades on equipment. A dull blade overworks the motor and can be unsafe.

- Generally, the cutting action of most tools is in a direction that is away from the body.

- Pay strict attention when operating any tool.

- Wear OSHA-approved safety glasses, footwear, and hard hat.

- Use proper gauge extension cords.

- Discard damaged electrical cords. Even a repaired cord is unsafe and can cause electrical shock.

- Adjust blade depths properly (usually ⅛ inch to ¼ inch more than the thickness of the stock that is being cut).

- Stabilize material or stock before cutting.

- Hold tools firmly and in a safe manner.

- Unplug tools before changing blades.

- Make sure the revolutions per minute (rpm) is built up before blades make contact with stock.

- Make sure that a saw base sits flat on the material being cut.

- Use push sticks to keep hands away from cutter heads and blades.

- Inspect trim and make sure there that no hidden nails or staples are present before making cuts.

- Use a backing board when cutting smaller pieces of trim on the miter saw. A backing board will help prevent smaller pieces from being launched from the saw (which can cause serious injury to anyone standing nearby).

- Finish nails sometimes curl out of wood due to knots and grain patterns. While using finish nailers, keep hands and fingers a safe distance away, just in case.

- Use safety railings on scaffolds and do not roll scaffolding over cords or hoses (doing so can cause scaffolding to tip over).

- Make sure ladders are secure before climbing up them.

- Keep workspaces clean.

- Do not remove or pin back guards on tools.

- Wear safety glasses when operating finish nailers or power tools.

- Wear hard hats when people are working above you.

- Wear proper clothing. Baggy clothing and unbuttoned cuffs can get caught in power equipment.

- Dispose of frayed electrical cords.

- Use machines for their intended purpose only.

Trim Carpentry

Introduction to Finish Carpentry

OBJECTIVES

After studying this chapter, you should be able to

- Discuss what is involved in being a finish carpenter
- Describe the materials used in finish carpentry
- Describe how to set up an ideal workstation
- Discuss the importance of a clean work environment
- Discuss on-the-job safety
- Explain the importance of understanding finish carpentry basics
- Discuss the general order of tasks and procedures in finish carpentry

WHAT IS FINISH CARPENTRY?

The finish carpenter is brought in during the final stages of construction, after all of the rough framing has been completed and after the **sheetrock** has been installed and has been taped, **bedded**, and textured. The finish carpenter's job is to install interior doors and to install molding wherever it is needed and requested. The job description and duties of the finish carpenter may vary slightly from region to region. For example, in some regions, the finish carpenter may not be expected to install cabinets or hardwood flooring. Those things may be assigned to another contractor.

Duties of Finish Carpenters

The word "construction" can bring to mind an endless number of images. Construction can mean excavation or steel work, or anything in between. "Carpenter" is not a very specific title either, considering the multitude of things that can fall under that title. When a person says, "I'm a finish carpenter," you know, or at least have a pretty good idea, what the person means by that. Perhaps images of grand staircases and delicate trim work come to mind.

Following is a list of the general duties of the finish carpenter. The finish carpenter is expected to

- Cut and install base molding, **wainscot**, picture molding, **chair rails**, and **crown molding**

- Install interior doors

- Case and trim windows

- Finish closets (shelves, shoeboxes, and hanging rods)

- Install cabinets and complete countertops

- Create ceiling beams

- Finish staircases

- Make and install columns

- Install hardwood flooring

- Build fireplace mantles

Materials Used by Finish Carpenters

There are more choices than ever when it comes to materials used for construction and finish carpentry. In an effort to conserve natural resources, companies are now manufacturing synthetic building materials. Consumers can now choose to use something other than solid wood products. These **alternative materials** are typically less costly than solid wood products and, unlike natural wood, manufactured materials are free from defects (e.g., knots, split grain, warps). Manufactured materials are straight and free of the many defects found in natural wood. Manufactured products are ideal when a paint finish is desired. Composite materials are now being made of plastic, polyvinyl chloride (PVC), polyethylene, polyurethane, fiberglass composite, polymer resins, fiberglass-reinforced gypsum, and the list goes on. Medium density fiberboard (**MDF**) is another popular alternative material currently being used. It is composed of compressed wood fibers and resin. It is cured under heat and pressure, which gives it an extremely hard surface. MDF is used for just about all of the typical moldings used in new home construction (e.g., crown molding, base molding, chair rail, door and window **casing**, etc.). MDF molding is preprimed and ready for paint. It is also available in sheets that are ideal for closet shelving. Interior doors constructed of MDF are being used in many homes today. Most building material centers (e.g., Home Depot, Lowe's) have catalogs and literature on different types of moldings and materials currently available. Some home builders and contractors keep a portfolio full of manufacturer catalogs (catalogs containing photographs, profiles, and dimensions) so that they can show homeowners or customers choices that are available to them regarding interior molding.

> **NOTE**
>
> Despite many different types of materials to choose from, the methods of installing these materials are basically the same and are covered in detail in the chapters to follow. ■

Finish carpenters also use a great deal of traditional materials, such as **stain-grade** and **paint-grade** wood. Stain-grade describes wood that is of good quality and suitable for staining. Paint-grade is a term used to describe material that is suitable for painting only. Interior molding is milled from both softwood and hardwood. Hardwood is used almost exclusively for projects requiring a stain finish; however, both softwood and hardwood can be stained. Composite materials, such as plastic and MDF, are used when a paint finish is desired.

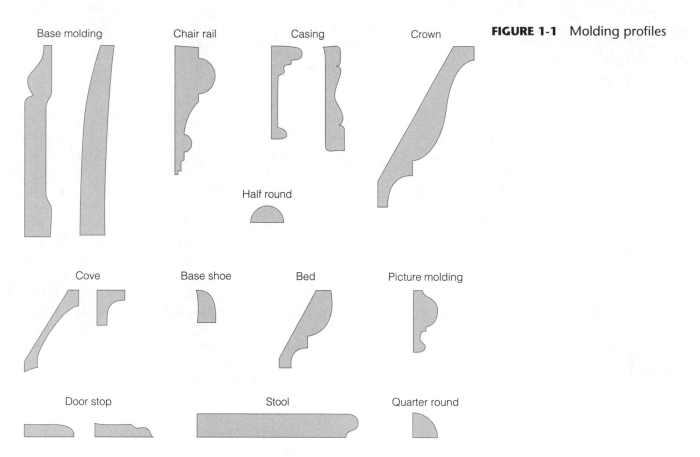

FIGURE 1-1 Molding profiles

Molding is sold by the linear foot and commonly comes in lengths of 7, 8, 10, 12, 14, and 16 feet. Seven-foot lengths are ideal for door casings because there is little waste. Moldings are available in many different **profiles** and **dimensions**. A profile is an end or side view and shows the molding's contour, height, and thickness (**Figure 1-1**). Dimension describes a material's height, length, and thickness. Dimensions are given in feet and inches and fractions of an inch.

Finger-jointed trim is made of short pieces of wood that are joined end-to-end to form one long unit. After gluing and joining, the long section is **milled** with a profile and afterward it is cut to standard trim lengths. Material is milled when it is sent through a machine to be planed, cut, or given a profile. One advantage to finger-jointed trim is that often it is much straighter than solid wood. Long sections of solid wood are sometimes bowed for various reasons; wood-grain pattern can cause a piece of wood to have a natural bend. Also, drying processes bring out a wood's natural tendency to bend. Finger-jointed trim is used only when a paint finish is desired; the reason for this is that joints are highly visible, there is no continuous grain pattern, and often the joined sections vary in shades of color. One section may be much lighter or darker than the piece joined next to it.

Flexible molding is made of a bendable polymer. It is used in areas that are curved. Some walls, especially in homes today, are constructed with a radius and the trim must follow the same curve. Using a flexible molding in such situations saves valuable time. Flexible molding is produced as crown molding, base molding, and window and door casing, as well as some others. Flexible molding can be cut using a **miter** saw. To make cutting flexible trim easier, lay it on a 2×4 or 2×6; doing so will add support so that the proper cut can be made. Instead of trying to cope flexible trim, leave the ends square and cope the pieces that are joining to it. **Coping** is cutting the end of trim in such a way that allows it to fit precisely over an adjoining piece of trim. Flexible trim can be fastened with the same fasteners and techniques used for any other type of molding.

Joints Used in Finish Carpentry

Figures 1-2 through **1-7** show some typical joints used in finish carpentry. **Tongue and groove** is commonly used with flooring, ceiling materials, and bead board paneling. One edge of the material is milled with a tongue and the other edge is milled with a groove for accepting the tongue. Tongue and groove are fastened

FIGURE 1-2   Tongue-and-groove joint

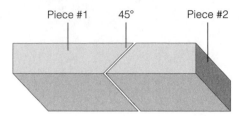

FIGURE 1-3   Scarf joint

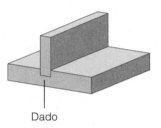

FIGURE 1-4   Dado (groove is made with a dado blade or straight fluting bit)

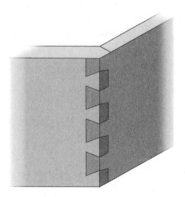

FIGURE 1-5   Dovetail joint

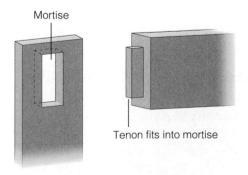

FIGURE 1-6   Mortise-and-tenon joint

FIGURE 1-7   Lap joint

with a method referred to as **blind-nailing**, a technique in which nails are driven into the tongue portion of the material (after which the nails are covered and hidden by the joining piece). Blind-nailing is covered thoroughly in later chapters.

Scarf joints are used to **splice**, or join, most interior moldings (**baseboard**, crown, cove, chair rail, etc.). When a section of molding is not long enough to go from one wall to another, two pieces have to be spliced together. Splicing, if done correctly, is not noticeable; the two pieces of molding will appear as one long piece. Splices and joints in general should always be glued and nailed over something solid, such as a wall stud or ceiling joist (scarf joints and splices are covered more in later chapters).

Dado joints are sometimes used in cabinet making and they are used on some shelving units. A dado is cut at a width slightly larger than the thickness of the material being used so that the material can fit precisely into the space.

Dovetail joints and **mortise-and-tenon joints** are some of the oldest joinery methods in history. They are used in some aspects of cabinetry and furniture making. They are very good joints but not always practical for reasons of efficiency and time management.

Rabbet joints can be found where the head jamb and side jamb meet on door and window jamb units. The rabbet is a shelf along the edge or across the end of material in which joining material can rest. The rabbet locks materials together. If materials were joined without using the rabbet, the pieces could possibly shift or twist with the passing of time (especially in the case of wood, as the material continues to dry even after installation). Like other joints, rabbets should be *glued* and *nailed*. Rabbets are used to create lap joints.

Glue and Sandpaper

Two allies of the finish carpenter are *glue* and *sandpaper*. These two things can make the difference between a great finish carpenter and a mediocre one. A joint or corner without glue will more than likely reopen after time. Six months to a year after the job has been finished, people may notice that the molding has opened at the joints and corners; this is easy to notice, especially when the paint has cracked and a dark line is present on the white high-gloss trim. Even if the joint was good and tight when it was completed, without glue it has a greater chance of opening while the house is acclimating and being lived in. Houses and structures shift and settle, even if only slightly, in the months after construction; framing lumber continues to dry and contract, and gravity pulls on the weight of all the materials, causing minor movements. Without glue, the finish work will show signs of this movement.

Wood glue will help joints and corners remain tight. Use good-quality glue, preferably one that is waterproof. Different regions of the country have different levels of humidity. Waterproof glue will not be affected by normal levels of humidity. Use glue on all corners, joints, and splices. Glue overrun and *squeeze-out* should be wiped off material surfaces with a damp cloth. Be particularly careful of getting glue on stain-grade surfaces. Wood will not accept stain in areas where glue has been allowed to dry, which will result in the finished product having a splotchy or bleached-out appearance.

Sandpaper ranges in grit from 40 (very coarse) to over 1,000 (very fine). Coarse sandpaper is used to remove material. Finish carpenters typically use sandpaper ranging from 100 to 220 grit. Grit from 100 to 150 is ideal for sanding joints, splices, and outside corners; any finer than that and the job becomes very time-consuming. The objective is to smooth the joining pieces of material without leaving scratches or scuff marks that will be visible after the final finish, or the paint. Painters typically use 150- through 220-grit sandpaper. A joint should be sanded smooth; otherwise, it will be highly visible after the finish (or paint) has been applied.

NOTE

Always sand with the grain. Never sand across the grain of wood. Sanding across the grain tears wood fibers, leaving what will show up as scratches after the material has been painted. Composite materials should be sanded in a similar fashion even though there is not a grain, sanding lengthwise and not across the material's width. ■

SETTING UP THE WORKSTATION

The work of the finish carpenter is made easier when a comfortable and efficient workstation is created. **Figure 1-8** shows one type of workstation. A pair of plastic folding sawhorses holds a 4-foot by 8-foot sheet of **plywood**, on which most of the tools used on the job can be positioned in a way that makes them easily accessible, saving valuable time. Time is money, and efficiency is everything in today's competitive market. Fumbling over tools and cords is unsafe, causes unnecessary headaches, and severely hinders production.

Having portable tools and a portable setup is important for finish carpenters. After the job has been completed, everything has to be transported to another location; therefore, stationary industrial tools are not at all practical. **Figure 1-9** shows some stationary tools. Obviously, these heavy-duty industrial tools would be too heavy to move on a regular basis. These

FIGURE 1-8 Workstation

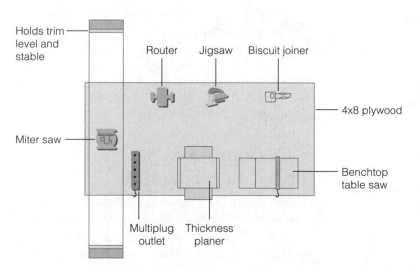

FIGURE 1-9   Stationary tools

types of tools are common in cabinet shops, which is why, as discussed in later chapters, cabinets are made elsewhere and then delivered to jobsites.

When looking for a place to set up a workstation, it is important to remember the length of the trim that will be cut. Trim usually comes in lengths anywhere from 7 feet to 16 feet, which means that much distance is needed on either side of the workstation. Also, the location must be out of the way of workers in other trades. Setting up out of their way will keep you from having to relocate.

MAINTAINING THE WORKSITE

A clean work environment is important for both safety and efficiency reasons. Scrap pieces of trim and piles of sawdust can lead to serious accidents, such as tripping while a power tool is in use. A messy workspace tends to break the worker's concentration by exuding a sense of chaos, slowing down production. A job runs more efficiently when things are kept clean and organized. Keeping a trashcan and a broom close by is a big help in keeping the area clean. A clean workspace reflects professionalism and could influence your client's opinion of you and your work.

Following are some general recommendations for maintaining the worksite:

- Clean throughout the day, and especially at the end of the day.

- Store materials in a way that is *organized* and prevents them from warping.

- Route extension cords in a way that keeps them from becoming tangled.

JOB SAFETY

Job safety is the responsibility of each individual. Big commercial jobs employ safety personnel who are responsible for seeing that everyone follows the safety rules. They cannot, however, be around all the time. It takes only one second of carelessness to cause an accident. Accidents occur in construction every day and leave people permanently disabled or worse. Fatalities are not uncommon in today's workplace, despite the rules and regulations of the Occupational Safety and Health Administration (**OSHA**) (see the section on OSHA later in this chapter).

The following safety tips are general recommendations that apply to all finish carpentry work.

  SAFETY TIPS

- Wear proper clothing. Baggy clothing and unbuttoned cuffs can get caught in power equipment.
- Wear approved safety glasses, footwear, and hard hat.
- Always wear approved safety glasses when operating power tools.
- Wear hard hats when people are working above you.
- Wear a dust mask when cutting or sanding, especially when working with MDF.
- Use tools and machines for their intended purpose only, and never remove or turn off their safety features.

General safety guidelines for electrical safety are provided later in this chapter.

The Occupational Safety and Health Administration

The main priority of OSHA is to protect the health and save the lives of American workers by preventing injuries. OSHA enforces safety regulations and issues heavy fines to employers who violate safety rules. The fines serve as a reminder to anyone who would put production ahead of safety.

OSHA inspectors cannot be everywhere, so it is up to individuals and companies to look out for their own best interests. Even doing the right things will not guarantee that your career in construction will be without incident. Other people's mistakes, such as those of careless coworkers, can result in injury. So, not only should you be aware of your own actions, but you should also be aware of the actions of those around you at all times. Following all safety recommendations will greatly reduce the chances of an accident occurring.

Electrical Safety

Electrical safety is a critical safety issue, since electricity is used on a daily basis in finish carpentry. When at all possible, plug extension cords into ground-fault circuit interrupter (**GFCI**) receptacles. These outlets are built to protect equipment as well as personnel.

The GFCI receptacle has a test button and a reset button on its face (**Figure 1-10**). These need to be tested periodically to make sure they are working properly. The GFCI outlet is designed to shut off when a ground or an imbalance of electricity has occurred, meaning that it should shut itself off before serious damage can occur to human or machine.

The following safety tips are some things to keep in mind at all times when using electricity.

SAFETY TIPS

- Use only GFCI receptacles.
- Test GFCI receptacles regularly—daily if possible, but no less than once a week.
- Use proper gauge extension cords.
- Never work around wet areas, and do not let extension cords run across or through wet areas.
- Replace frayed or damaged electrical and extension cords. Repaired cords are unsafe and can cause electrical shock.
- Unplug all tools and extension cords at the end of each work day.

General safety tips to keep in mind when using tools are covered in Chapter 2.

Job Safety Basics

Besides costing employers money, workplace accidents also result in

- Workers' compensation costs
- Costs to insurance companies
- Lost work time
- Liability costs
- Reduced production
- Physical pain and suffering

The methods covered in the chapters that follow are examples of how to correctly perform specific tasks in the finish carpentry trade. The methods discussed are not the only ways of doing things, but they are methods that are widely accepted in the industry. Many experienced carpenters develop their own ways of doing things; their methods may vary from those described in this book. With time and experience, you may develop your own style and method of doing certain things as well.

The chapters that follow provide a basic foundation for understanding how to complete the tasks of finish carpentry. Jobs may be accomplished in many ways; ultimately, it is up to individuals to learn what works best for them in terms of producing quality work while still ensuring a safe workplace.

FIGURE 1-10 GFCI receptacle

FINISH CARPENTRY BASICS

The following chapters are intended to assist in teaching the basic skills needed for understanding and accomplishing every aspect of finish carpentry. It is impossible to cover every problem and situation encountered during a career in finish carpentry. But with the knowledge and understanding of finish carpentry basics, you can overcome the challenges in almost any situation.

The rest of this section covers several general areas of knowledge in finish carpentry basics: the order in which tasks are done, definitions of terms, and general procedures. Use this material as a resource for understanding finish carpentry basics; and as you read the later chapters, refer back to this chapter as necessary to clarify your understanding of general tasks, terms, and procedures.

Order in Which Tasks Are Performed

The order in which tasks are performed may vary depending on the individual carpenter. Many of the finish carpentry tasks can be done independently of one another, as they do not interfere with other tasks. For example, crown molding can be completed first because it does not interfere with any other task. Certain tasks, however, must be completed before other tasks can begin. For example, doors must be installed and trimmed so that the base molding can tie into the door casing, and usually closets are not completed until the base molding is installed (the bottom of some shoeboxes rests on top of the base molding).

Following is a list showing the order in which finish carpentry tasks typically are completed in a new custom home. Generally, the finish carpenter proceeds with installation in this order:

1. Interior doors and door casings
2. Base molding
3. Windows (jamb extensions, casings, **stools**, and **aprons**)
4. Wainscot, picture molding, and chair rails
5. Closets
6. Crown molding
7. Ceiling beams
8. Stairs
9. Columns
10. Cabinets
11. Hardwood flooring (followed by the installation of **shoe molding**)

Making a Punch List

Making a list, often referred to as a **punch list**, toward the end of a project is very beneficial. A punch list is simply a list of remaining tasks. Making a thorough and complete list will help you estimate the time it will take to complete the job. It will also help ensure that nothing is forgotten. Sometimes, in the chaos and rush of things, it is easy to overlook some minor detail. A punch list helps keep things on a professional level by eliminating the chance of being called back to a job because of an overlooked detail. The list can consist of just a few items or it may list two pages of tasks that need to be done. As the tasks are completed, simply mark them off the list. The following is an example of a punch list:

■ Cut and install closet rods.

■ Finish baseboard in bathroom and around kitchen cabinets.

■ Trim beside fireplace.

■ Adjust door in master bedroom.

■ Install chair rail in dining area.

General Finish Carpentry Procedures

The following general procedures should be followed in all aspects of finish carpentry:

■ Use a finish hammer that has a smooth face. Avoid making contact with the surface of the wood by using a **nail set** to **countersink** nails.

■ Nails should be kept at least ¾ inch away from the end of trim to avoid splitting the wood.

■ If you are using hand-driven nails, the nail point should be blunted before use.

■ Corners and joints should be glued and nailed.

■ Sand joints and outside corners with fine-grit sandpaper (100 to 220 grit).

■ Slightly round any sharp edges (e.g., window and doorjamb edges) using fine-grit sandpaper (100 to 220 grit). Sharp edges are easily damaged and do not accept paint as well as slightly rounded edges.

- Always sand with the grain of the wood—never sand across the grain. Sanding across the grain tears and damages wood fibers, showing up as scuff marks after the final finishing.

- Sand or erase pencil marks off wood that will be stained. Pencil marks will be visible in the finished product.

- Exposed glue should be sanded away completely; otherwise, wood will not accept stain in areas where glue has been allowed to dry.

- Newly applied glue should be wiped off with a damp rag.

- Make sure joints fit properly and are tight. Do not leave loose or open joints.

> **NOTE**
>
> The edges of **veneer** or **plywood** should be covered with a molding of some type and should never be left exposed. The edge of veneer is not suitable for shaping with a router. ■

SUMMARY

- The finish carpenter is brought in near the finishing stages of construction.
- The finish carpenter's job is to install interior doors and molding after the rough framing is completed and sheetrock has been installed.
- Knowledge and understanding of finish carpentry basics enable one to overcome the challenges of almost any situation.
- Having portable tools and a portable setup is important because it enables the finish carpenter to easily transport equipment from one worksite to another.
- An efficient workstation saves time and money.
- Worksites should be kept clean at all times for both safety and efficiency reasons.

- Job safety is the responsibility of each individual worker.
- OSHA was established to protect the safety and well-being of American workers.
- Safety features on all tools and machines should be used at all times.
- Tools and machines should be used only for their intended purposes.
- Wear approved safety glasses and footwear when operating power tools, and a hard hat when people are working above you.
- Electrical safety is a critical issue because finish carpenters use power tools daily.
- GFCI receptacles should be tested frequently, at least once a week.

KEY TERMS

alternative materials Materials other than wood that are being widely used in the construction industry (e.g., composites, recycled materials, plastics). Synthetic materials are typically less costly than solid wood products and using them helps conserve natural resources.

aprons Trim members installed beneath window stools.

baseboard Trim that covers the area where floor and wall meet.

bedded Joint compound has been applied to taped joint of drywall.

blind-nailing A method of nailing in which the nail is hidden from sight, usually near the tongue of the tongue-and-groove material.

casing Trim around doors, windows, and cased openings.

chair rails Trim typically installed at a height of 3 feet off the floor, designed to keep chairs from hitting walls. Chair rails are sometimes used for the cap of wainscoting.

coping Cutting the end of trim so that it will fit precisely over the adjoining piece of molding.

countersink To drive the head of a nail or screw in below the surface of the material. A nail or screw is countersunk so that wood filler can be applied and sanded flush with the surface of the material.

crown molding Molding that is installed where wall and ceiling meet; rests at an angle between these two structural components; also known as cornice.

dado A rectangular-shaped groove cut into wood; a joining technique generally used with shelving; a slot cut across the grain of the wood.

dimensions A material's height, length, and thickness. Dimensions are given in feet and inches and fractions of an inch.

dovetail joints Some of the oldest joinery methods in history. They are still used in some aspects of cabinetry and furniture making. They are very good joints but are not always practical for reasons of efficiency and time management.

finger-jointed Trim produced by a joining method in which small pieces of wood are joined to form one long piece, after which that piece is milled into a molding, such as base or crown, that is suitable only for painting.

flexible molding Molding made of a bendable polymer; used in areas that are curved. Some walls, especially in homes today, are constructed with a radius and the trim must follow the same curve. Using a flexible molding in such situations saves valuable time. Flexible molding is produced as crown molding, base molding, and window and door casing, as well as some others. Flexible molding can be cut using a **miter** saw.

GFCI Ground fault circuit interrupter. The GFCI outlet is designed to shut off when a ground or an imbalance of electricity has occurred, meaning that it should shut itself off before serious damage can occur to human or machine.

MDF Medium density fiberboard. A material composed of wood fiber and resin; used for making different types of molding (e.g., baseboard, crown, casing, chair rail, etc.); an alternative material. It is also available in 4-foot × 8-foot × ¾-inch and 1-inch × 12-inch × 16-foot sheets.

milled Sent through a machine to be planed, cut, or given a profile.

miter An angled cut that allows two pieces of trim to join at a corner or splice.

mortise-and-tenon joints Older joinery methods; still used in some aspects of cabinetry and furniture making. They are very good joints but are not always practical for reasons of efficiency and time management.

nail set Tool used for countersinking nails.

OSHA Occupational Safety and Health Administration; their main priority is to protect the health and save the lives of American workers by preventing injuries.

paint-grade Materials that are suitable for painting. Composite materials, such as plastic and MDF, are used when a paint finish is desired.

plywood A product made up of multiple layers of wood; commonly available in 4-foot × 8-foot sheets.

profiles End or side views; show the molding's contour, height, and thickness.

punch list A list of remaining tasks.

rabbet Two-sided cut that runs down the length of a board or across the end of a board (e.g., shelf where the head jamb of a door or window rests on the side jamb).

scarf joints Joints where two pieces meet at a 45-degree angle; two pieces of material are spliced together with a scarf joint.

sheetrock Often used as a generic term for describing drywall or gypsum wallboard; "Sheetrock" is actually a brand of drywall fiberglass-reinforced gypsum encased in heavy paper. It is applied to interior walls and ceilings; the edges are tapered so that joint compound can be applied to conceal joints; usually finished with a plaster-based texture and then painted.

shoe molding Small trim member mounted at the base of baseboard; rests against both the base and the floor; used to cover small gaps between baseboard and flooring.

splice To join two pieces of molding together using miter cuts, usually consisting of 45-degree cuts; where two pieces of trim join at the ends. Two pieces of trim often have to be spliced together if one piece is not long enough to go from wall to wall.

stain-grade Wood that is of good quality and suitable for staining. Interior molding is milled from both softwood and hardwood. Hardwood is used almost exclusively for projects requiring a stain finish; however, both softwood and hardwood can be stained.

stools The bottom portions of window trim.

tongue and groove Commonly used with flooring, ceiling materials, and bead board paneling. One edge of the material is milled with a tongue and the other edge is milled with a groove for accepting the tongue.

veneer A thin layer of wood; thin layers of wood glued together to make plywood.

wainscot A wall treatment covering the bottom portion of a wall, typically finished at a height of 32 or 36 inches, and often trimmed at the top with cap rail or chair rail.

REVIEW QUESTIONS

1. What does OSHA stand for?
2. What is the primary function of OSHA?
3. List three tasks that are the responsibility of the finish carpenter.
4. Explain the importance of having an efficient workstation.
5. Why do most finish carpenters use portable tools instead of stationary tools?
6. Is it okay to bypass the safety features on certain tools?
7. Is it okay to disregard safety rules when OSHA is not around? Why or why not?
8. Is there one universally accepted method of doing things that is shared by all finish carpenters?
9. Why does the order in which a finish carpenter completes the tasks matter? Give an example.
10. Is finger-jointed trim suitable for staining?
11. Name two methods for joining wood.
12. Why should nails and screws be countersunk?
13. For best results, should wood be sanded across the grain or with the grain?
14. How often should GFCI receptacles be tested?
15. What effect does glue have on wood that is going to be stained?

CHAPTER 2

Tools Used in Finish Carpentry

OBJECTIVES

After studying this chapter, you should be able to

- List techniques that workers use to ensure job safety while operating tools
- Identify the two main categories of tools used in finish carpentry
- Describe tools used in finish carpentry and their functions
- Discuss tool care and maintenance

INTRODUCTION

A wide variety of quality hand and power tools are available on the market today. Experience and familiarity with different brands of tools will help the finish carpenter in determining which tools will allow safe job performance along with the highest-quality work that is possible. An entire book could be written on tools and their many uses and functions. This chapter provides an overview of the tools used in the finish carpentry trade and their functions. ∎

TOOL SAFETY

Safety cannot be stressed too much where tools are concerned. Follow manufacturers' instructions and guidelines at all times. Be familiar with all safety guidelines before attempting to operate any tool.

Following is a list of some safety tips to keep in mind *before, during,* and *after* any work is performed.

 SAFETY TIPS

- Use tools and machines for their intended purpose only.
- Pay strict attention when operating any tool.
- Hold tools firmly and in a safe manner.
- Make sure that the equipment switch is in the OFF position when plugging the cord into a power outlet, and make sure all tools and cords are grounded.
- Leave guards and safety features intact, and check them on a regular basis to make sure they are in proper working order.
- Use safety railings on scaffolds, and do not roll scaffolding over cords or hoses (doing so can cause scaffolding to tip over).
- Make sure ladders are secure before climbing up them.

In addition, keep the following safety tips in mind when using cutting tools such as saws and routers.

 SAFETY TIPS

- Keep blades on equipment sharp. A dull blade overworks the motor and can be unsafe.
- Unplug tools before changing blades or performing any other kind of work on them.
- Do not remove or pin back guards on tools.
- Inspect trim and make sure there are no hidden nails or staples present before making cuts.
- Stabilize material or stock before cutting, and make sure the saw base sits flat on the material being cut.
- Properly adjust blade depths (usually ⅛ to ¼ inch more than the thickness of the stock that is being cut).
- Make sure the rpm is built up before blades make contact with stock.

(continued)

- Cut material so that the cutting action of the tool is away from the body.
- Use a backing board when cutting smaller pieces of trim on the miter saw. A backing board will help prevent smaller pieces from being launched from the saw (which can cause serious injury to anyone standing nearby).
- Use push sticks to keep hands away from cutter heads and blades.
- Do not overwork a machine by cutting material at a faster rate than it can handle.

Individual tool safety is covered in more depth later in this chapter and throughout the rest of the textbook. Pay particular attention to the safety tips and cautions that appear in later chapters. As stated at the beginning of this section, safety cannot be stressed too much when working with tools.

TYPES OF TOOLS

The tools used in finish carpentry can be broken down into two broad categories: hand tools and power tools. **Hand tools** are tools that do not require electricity, while **power tools** are those that do require electricity or some alternate source of power. There are tools that operate using *gas* (e.g., finish nailers, framing nailers). *Battery-powered tools* are very popular in construction; they are practical because they can be used without the hindrance of an extension cord. For example, a battery-powered drill can be carried from room to room without having to reroute a long extension cord. Battery-operated tools are very efficient in that they save time and energy. *Pneumatic tools* are those that operate using compressed air (e.g., finish nailers, framing nailers, and flooring nailers). Gas, battery-powered, and pneumatic tools fall into the category of power tools because, unlike hand tools, they require an alternate source of power.

Hand Tools

Hand tools are used extensively in finish carpentry. They should be used in a safe manner. Improper use of hand tools can cause serious injury. Following is a brief description of various hand tools and their uses in finish carpentry.

Measuring, Leveling, and Framing Instruments

Tape measure. Tape measures (**Figure 2-1**) are used for measuring materials at exact lengths. Tape measures are available in many different lengths. Finish carpenters typically use 20- to 30-foot–long tape measures that are 1 to 1½ inches in width. Tape measures have a moveable hook that slides in either direction, ensuring that both inside and outside measurements are accurate. If a tape measure is dropped, check to make sure the hook has not been bent. A tape measure with a bent hook will not yield accurate measurements. Extremely accurate measurements can be obtained by **cutting an inch** (also referred to as burning an inch). To measure by cutting an inch, begin the measurement directly on the 1-inch mark, and mark the desired measurement. Do not forget to add the *cut inch* to the overall measurement (e.g., instead of marking 12 inches, you would place the mark on 13 inches to make up for the inch that was cut).

Chalk line. The chalk line (**Figure 2-2**) is used for snapping straight lines. Blue chalk should be used on walls because red chalk is permanent and bleeds through paint. The line and the chalk should be kept dry. Before snapping a line, remove excess chalk by holding the line tight and lightly snapping it once. Removing the excess chalk will leave a narrower and more defined line on the finished piece.

Spirit level. Levels (**Figure 2-3**) commonly come in lengths of 2, 4, and 6 feet. Torpedo levels are about 9 inches in length. Levels are used to see if things (e.g., cabinets, interior doors, shelves) are level and plumb. **Level** means flat, perfectly horizontal, in relation to gravity. **Plumb** means perpendicular, straight up and down, at a right angle to something that is level. Most levels are equipped with two or three transparent vials. The vials are filled with mineral spirits (to help prevent freezing in cold temperatures), which is where the name *spirit level* comes from. When the air bubble inside the vial is resting between the lines, the level is plumb, or level. Dropping the level can shift the vial alignment, making the level inaccurate.

T-bevel. A T-bevel (**Figure 2-4**) helps in determining angles and in marking angles, dovetails, and so on. It has

FIGURE 2-2 Chalk line *(Courtesy of Floyd Vogt)*

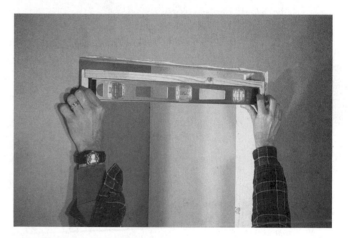

FIGURE 2-3 Spirit level

FIGURE 2-1 Tape measure

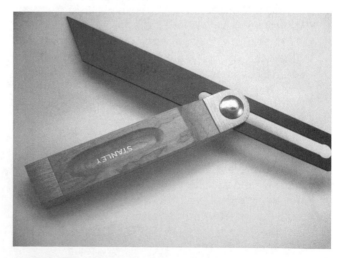

FIGURE 2-4 T-bevel

FIGURE 2-5   Compass

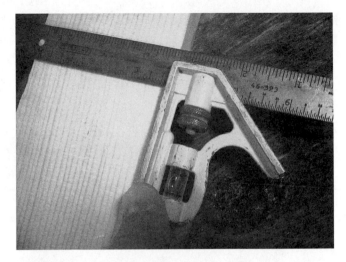

FIGURE 2-7   Combination square

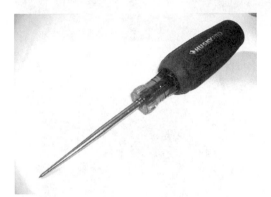

FIGURE 2-6   Scratch awl

FIGURE 2-8   Speed square

an adjustable blade. Once the correct setting is obtained, tightening the setscrew keeps the blade from moving.

Compass. A compass (**Figure 2-5**) is used for laying out arcs and radial designs. One arm of the compass has a metal point while the other arm holds a pencil or a sharpened piece of lead. The point remains in position while the lead marks the arc or radius. *Trammel points* work similar to a compass and are used for making larger arcs and radiuses.

Scratch awl. A scratch awl (**Figure 2-6**) has a sharp point and is used for scribing material, leaving a thin mark. A pencil leaves a line of varying thickness, depending on how sharp the lead is. The line left by a scratch awl is consistent. A small scratch awl can be found in the head of most combination squares.

Combination square. A combination square (**Figure 2-7**) is an adjustable square that combines a 90-degree angle, a 45-degree angle, a spirit level, and an awl. It can be used as a gauging instrument. The adjustable blade is ideal when it comes to tasks such as marking reveals around door and window jambs.

Speed square. The speed square (**Figure 2-8**) is used for marking square lines and 45-degree angles. It is also used for checking that various things are square. The speed square is marked with degrees and a scale of pitches. The square can be used to mark a board at any desired pitch or degree. It is also used as an aid in making straight cuts with the circular saw; this is done by holding the square firmly as the **shoe**, or base, of the saw moves along the edge of the square, making a square cut.

Framing square. The framing square (**Figure 2-9**) is used for marking shelves in closets, checking that closets are square, marking stringers, and general layout. The short part of the square is referred to as the tongue. The longer part of the square is referred to as the blade, or body. Aluminum framing squares are lightweight and do not rust as steel squares do. Stair gauges can be fastened to the framing square so that accuracy is maintained while laying out stair stringers. One gauge is fastened to the tongue of the square while the other one is fastened to the blade, making it possible to maintain a consistent rise and run.

FIGURE 2-9   Framing square

FIGURE 2-10   Trim bar

Trim bar. The trim bar (**Figure 2-10**) is used mainly for pulling trim tightly into a cope, such as where the base molding meets in a corner. The trim bar is used to hold pressure against the piece of base being nailed; without it, the force of impact from the finish nailer may cause the molding to kick back, creating a gap between the two joining pieces.

Saws

Coping saw. A coping saw (**Figure 2-11**) is used for coping trim, which is cutting the trim's profile so that two pieces of trim will meet at an inside corner with minimal gap. The blade of the coping saw is thin so that it can be used to make intricate cuts in wood. Most people prefer that the teeth of the blade angle back toward the handle. This gives the user more control as the cuts are made while the saw is pulled toward the body. The edges of coped trim can be lightly

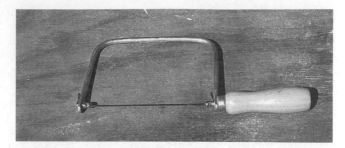

FIGURE 2-11   Coping saw

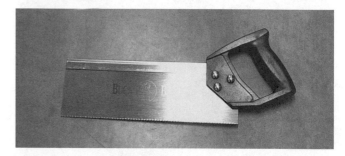

FIGURE 2-12   Backsaw

FIGURE 2-13   Dovetail saw

sanded, using fine-grit sandpaper, to smooth the contour. Coping saw blades are available with different pitches, such as fine, medium, and coarse. Dull blades should be discarded and replaced with sharp ones.

Backsaw. A backsaw (**Figure 2-12**) has a thin blade that is reinforced with a steel strip along the top. The blade produces fine cuts. The saw is good for making miter cuts and cutting dovetails.

Dovetail saw. A dovetail saw (**Figure 2-13**) produces a fine cut and is used primarily for cutting dovetails.

Another important finishing saw is the *Japanese Dozuki.* It is very sharp and produces a fine cut. It cuts on the pull stroke and cuts with the grain just as easily as it would cutting across the grain. It is good for flush cutting (e.g., cutting a wood plug flush with the surface that surrounds the plug).

Hammers and Chisels

Hammers. Hammers (**Figure 2-14**) are used for driving nails. The finish carpenter uses a smooth-faced hammer so as not to mar wood surfaces. (The head of the hammer should never make contact with anything

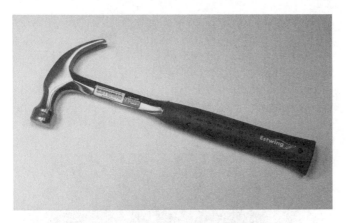

FIGURE 2-14 Hammer *(Courtesy of Floyd Vogt)*

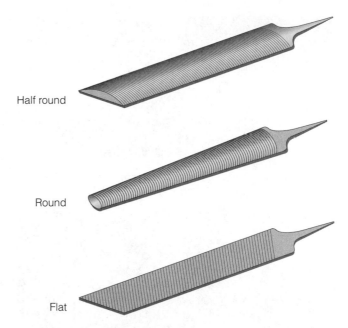

FIGURE 2-16 Wood files and rasps

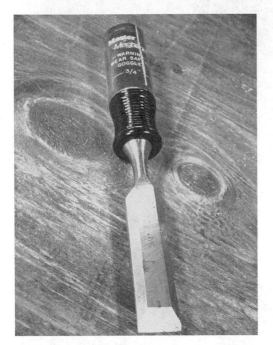

FIGURE 2-15 Chisel

FIGURE 2-17 Hand planes

other than a nail, nail set, or beater block.) Hammers are available in different head weights, ranging from 7 to 32 ounces. Finish carpenters typically use 16-ounce hammers. Framing carpenters usually use heavier 20-ounce hammers.

Chisels. Wood chisels (**Figure 2-15**) are used for mortising and removing excess wood, leaving a clean, smooth edge. Chisels should be sharpened regularly to prevent the blades from becoming dull. A dull cutting edge damages wood instead of leaving a clean smooth edge. Grinding wheels and sharpening stones are used in keeping chisels sharp.

Sanding and Planing Instruments

Wood files and rasps. Wood files and rasps (**Figure 2-16**) come in flat, round, and half-round shapes. They are used for shaping wood and removing excess wood, such as from a coped piece of trim.

Hand planes. Hand planes (**Figure 2-17**) are operated manually. They are designed to shave and smooth wood surfaces. Planes come in many different types and sizes. Some of the lesser-used ones today, due to more efficient alternatives, are the rabbetting plane and the router plane. Still being commonly used are the block plane and the jack plane. Blades on planes must be kept razor-sharp in order for the plane to work properly. The plane is used to shave the wood, gradually taking it down to the desired thickness.

Clamps, Nails, and Fasteners

Wood clamps. Wood clamps (**Figure 2-18**) are used for securing material. They lock material tightly into

FIGURE 2-18   Wood clamps

FIGURE 2-19   Nail sets

FIGURE 2-20   Finish nails, brad nails, and trim screws

position while work is being performed (e.g., routing, cutting, sanding), and they are used to hold two glued pieces together. Clamps should not be allowed to damage wood. If necessary, a block should be placed between the clamp and the wood surface to protect the wood.

Nail sets. Nail sets (**Figure 2-19**) are used for countersinking nails. Nails should be driven in below the surface, or countersunk, so that the hole can be filled with wood filler and sanded smooth. Nail sets come in different sizes. The size of the nail head determines which nail set should be used. Finish nails are set using the small nail set.

Fasteners: finish nails, brad nails, and trim screws. Fasteners used in finish work, such as finish nails, brad nails, and trim screws, are usually characterized by having a small head (**Figure 2-20**). A small head does minimal damage to trim.

Nails should be countersunk so that the head is below the surface of the wood. A nail set is used to countersink nails (**Figures 2-21a** and **2-21b**). A painting contractor usually fills the nail hole with a wood filler suitable for painting or staining, and the putty is then sanded flush with the surface of the wood using fine-grit sandpaper.

(a)

(b)

FIGURE 2-21   A nail set is used to countersink nails.

Trim screws, as well as all other types of screws, should be predrilled (**Figure 2-22a**) and countersunk (**Figure 2-22b**). **Predrilling** provides a path for the nail or screw to follow and keeps the wood from splitting. Ideally the head of the screw should be at a finished depth of $1/32$ inch to $1/16$ inch (**Figures 2-22c** and **2-22d**). A hole any deeper than that becomes difficult to fill with wood filler.

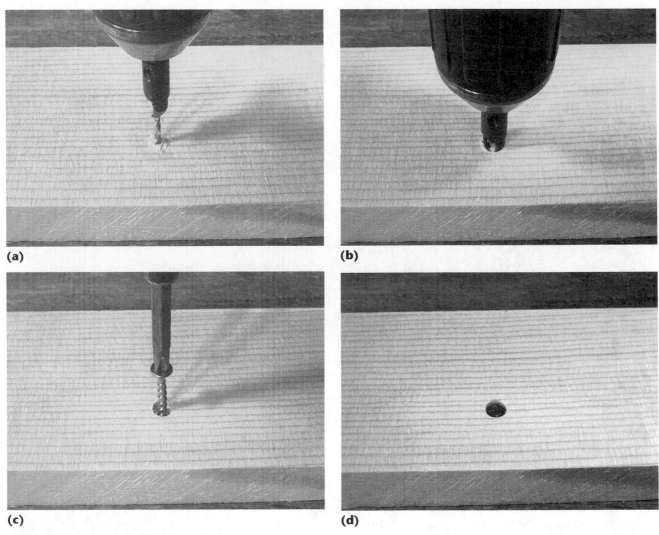

(a) **(b)** **(c)** **(d)**

FIGURE 2-22 Predrilling and countersinking a screw

Additional Hand Tools and Equipment

Other tools that may be necessary for the finish carpenter are *utility knife, pliers, cutout tool, crescent wrench, laminate trimmer,* and *pencils* and *pencil sharpener.* Keeping a sharp pencil is a must for the finish carpenter. Dull pencils leave a line that is too wide for making accurate cuts; some carpenters use the scratch awl (listed under Hand Tools), because it leaves a line of consistent thickness.

Commonly used smaller hand tools can be carried around at all times in a *tool pouch.* The market offers a wide variety of tool pouches. Some finish carpenters prefer tool pouches that have suspenders attached or built into them; this takes stress off the back as the shoulders help carry some of the weight. *Kneepads* are useful in providing a cushion between the knees and floor, especially when performing a task such as measuring and nailing base molding.

Other equipment used by finish carpenters includes *ladders of various sizes, air hoses, sawhorses,* and *heavy-gauge extension cords.* Because small areas, such as closets, normally have no windows and are dark, *halogen work lights* quite often come in handy on jobsites.

Power Tools

Power tools are used by finish carpenters daily, so practicing safety while using them is critical. Pay close attention to the safety tips and cautions in this section.

Measuring and Leveling Instruments

Laser level. The laser level (**Figure 2-23**) is used for marking level lines for cabinets, wainscot, closet shelves, and so on. Some laser levels place a level reference line on everything 360 degrees around the level. Most laser levels are self-leveling. This

FIGURE 2-23   Laser level

feature cuts down on the time it takes to set up the instrument.

Saws

Table saw. The table saw (**Figure 2-24a**) has an adjustable **rip fence** and is used for ripping stock to an exact width. The rip fence slides on a track and is locked into place once the desired setting is achieved. Most carpenters check the rip setting (**Figure 2-24b**) by measuring from the rip fence to the inside part of the blade. (Measurements are taken at both the front and the back of the blade to make sure the fence is square in relation to the blade.) Material cannot be cut properly if the fence is not perfectly parallel to the blade. The height (or depth) of the blade should be adjusted according to the thickness of the material being cut (**Figure 2-24c**). Usually the blade is set about ¼ inch higher than the thickness of the material (or to the bottom of the kerfs on the blade); this allows the blade to cool while cutting. The table saw also can be adjusted to cut angles of various degrees, usually from 0 to 45 degrees. Dull saw blades should never be used.

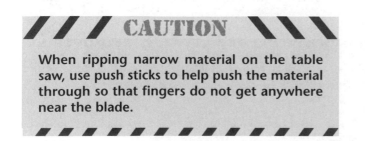

CAUTION

When ripping narrow material on the table saw, use push sticks to help push the material through so that fingers do not get anywhere near the blade.

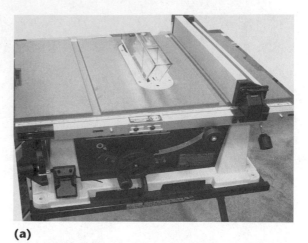

(a)

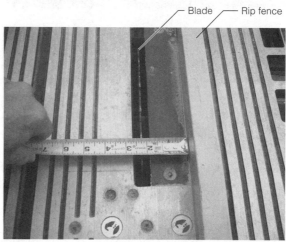

Blade   Rip fence

(b)

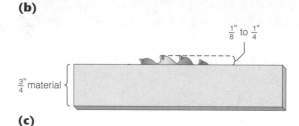

$\frac{1}{8}"$ to $\frac{1}{4}"$

$\frac{3}{4}"$ material

(c)

FIGURE 2-24   (a) Table saw; (b) Checking the rip width (the safety has been removed for visual purposes); (c) Height (or depth) of blade should be adjusted according to thickness of material.

Portable circular saw. The circular saw (**Figure 2-25**) is used for ripping and crosscutting stock, leaving semirough edges. Circular saws can be adjusted to make bevel cuts, and the depth of cut also can be adjusted. The blade depth should be set about ¼ inch deeper than the thickness of the material being cut.

Small pieces of wood can sometimes become wedged between the guard and the housing, causing the blade guard to malfunction. The blade guard should be inspected regularly to ensure that it is in working order. The portable circular saw is sometimes called a *skill saw* because it was first made by the company Skil in 1926.

FIGURE 2-25   Portable circular saw

FIGURE 2-26   Miter saw

FIGURE 2-27   Jigsaw

FIGURE 2-28   Drill

ther direction for making beveled cuts. Different types of blades can be used for cutting various types of material.

Drills and Routers

Drills. Drills (**Figure 2-28**) use interchangeable drill bits and can be equipped with tips for driving screws. Most drills are variable speed (speed is adjusted and controlled by depressing the trigger) and have a forward and reverse. Cordless drills use rechargeable batteries.

/// CAUTION \\\

Saws sometimes kick back when material binds on the blade. Hold saws with a firm grip. Always inspect material before cutting to make sure there are no hidden nails or staples present.

Miter saw. A miter saw (**Figure 2-26**) is used for cutting materials to proper length. It is capable of cutting at various angles, which are gauged by a scale of degrees that comes standard on any miter saw. Some miter saws are capable of making compound miter cuts as well as miter cuts. The material rests stationary on the saw base, while the head pivots downward to cut the material.

Saber saw. This is often referred to as a *jigsaw*. The saber saw (**Figure 2-27**) is used for making irregular and radial cuts. The shoe of the jigsaw can be tilted in ei-

/// CAUTION \\\

When using any type of power equipment, always make sure the rpm is built up before blades make contact with stock. A drill is the only exception to this rule. When you are using a drill, the end of the drill bit should be placed on the material at the exact location where the hole is to be drilled, and then the drill can be engaged.

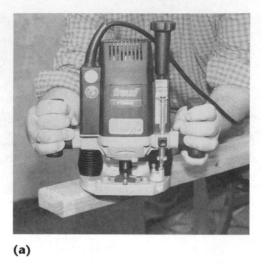

(a)

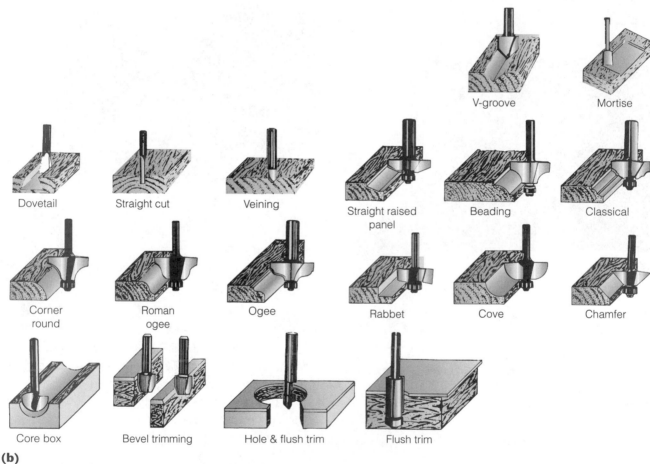

(b)

FIGURE 2-29   Router *(a. Courtesy of Floyd Vogt and b. Courtesy of Stanley Tools)*

Routers. Routers (**Figure 2-29**) use interchangeable bits that can cut decorative profiles in wood. The base of the router is adjustable to achieve different cutting depths. The rpm (revolutions per minute) of the router bit must be built up before making contact with the wood. Cutting too fast with the router will damage the wood. Cut slowly so that the motor does not bog down. For best results, use carbide-tipped router bits that are equipped with ball bearings. **Chatter** occurs when dull router bits are used or the when the router is overworked. Chatter consists of ripples in the wood that are left behind when the router bit vibrates against the wood.

Sanding and Planing Instruments

Palm sander, belt sander, and orbital sander. Electric sanders (**Figure 2-30**) use different grits of sandpaper, ranging from very fine to coarse. Sanders are used for smoothing wood surfaces. Finish carpenters typically

(a)

(b)

(c)

FIGURE 2-30 (a) Palm sander (b) Belt sander (c) Orbital sander

FIGURE 2-31 Thickness planer

use sandpaper ranging from 100 to 220 grit. Sandpaper ranges from 40 grit (coarse) to over 1,000 grit (very fine). Always wear a dust mask while sanding to prevent respiratory problems.

Thickness planer. The thickness planer (**Figure 2-31**) is used to shave wood down to an exact dimension. Most carpenters rip the stock slightly larger than what is needed and plane it down to the exact dimension. In this way, the planer also removes saw marks and leaves a clean, smooth edge. Do not try to shave too much at once. Most planers perform best when adjusted to remove $\frac{1}{32}$ to $\frac{1}{16}$ inch of surface. Make several passes if necessary to achieve the desired thickness.

//// **CAUTION** \\\\

Make sure wood is free of nails and staples before sending it through the planer.

Jointer. A jointer (**Figure 2-32**) is used to true and smooth wood edges. A crown or bow can be removed from a board using a jointer. The jointer uses a cutter head, similar to that of a planer, to shave off board edges.

Nailers and Joiners

Although *hand-driven* nails (nails that are installed using a hammer) can be used to fasten trim work, pneumatic finish nailers are preferred in today's

(a)

(b)

FIGURE 2-32   Jointer

competitive market. Pneumatic nailers are practical and more efficient than using hand-driven nails. For small tasks, hand-driven nails are sometimes used (such as in a situation when there is not much nailing to be done and it is not practical to get out an air compressor, air hose, and extension cord). When using hand-driven finish nails, it sometimes helps to drill a **pilot hole** first. A pilot hole, one that is slightly smaller than the nail being used, will guide the nail and will prevent the wood from splitting. Blunting the head of the finish nail will also help keep the wood, or other material, from splitting. Nails, whether hand-driven or from a pneumatic nailer, should be countersunk so that wood filler can later

be applied to cover the nail. Countersinking is done using a nail set (see section on hand tools in Chapter 1). Hand-driven nails are available in different sizes. The heads of finish nails are small so that a minimal amount of damage is done to the trim of material being nailed.

Finish Nailers. Finish nailers are widely used in trim work. Pneumatic finish nailers (**Figure 2-33**) use compressed air to drive finish nails of various lengths and gauges. Gas- or battery-powered nailers are also available. Finish nailers are used to fasten trim of all types. The nailer can be actuated once the safety device is fully depressed. The safety device is depressed when the nose of the nailer is pressed against the trim that is to be nailed.

When the pressure is adjusted correctly, nails from the finish nailer are driven in and are countersunk automatically. Occasionally nails fail to countersink, in which case a nail set must be used. The head of a hammer leaves dents and visibly damages trim, so it should never make contact with the surface of trim.

> ///// **CAUTION** \\\\\
>
> **Nails exit finish nailers at a high velocity and can ricochet off surfaces. Nails also can curl out of the wood due to knots or grain patterns. Always wear safety glasses and keep fingers a safe distance away from the nailing area. Never point finish nailers toward yourself or others.**

Pin nailers and brad nailers. Pin and brad nailers (**Figure 2-34**) fasten trim using smaller-gauge finish nails. Smaller nails are less likely to damage and split wood. Most pneumatic nailers require oil on a daily basis. Follow the manufacturer's suggestions regarding the type of oil to use and frequency of use.

Biscuit joiner. The biscuit joiner (**Figure 2-35**) is used for joining wood. A slot is cut in both pieces that are to be joined, and then a thin piece of wood called a *biscuit* is glued and inserted into the slot. The two pieces are clamped together and allowed to sit until the glue has had time to dry.

Air Compressors

Air compressors (**Figure 2-36**) supply pneumatic nailers and other tools with compressed air that is

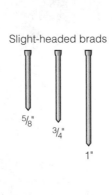

Finish nails

1" (2D)   1¼" (3D)   1½" (4D)   1¾" (5D)   2" (6D)

Slight-headed brads

5/8"   3/4"   1"

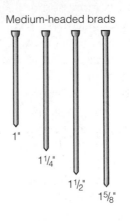

Medium-headed brads

1"   1¼"   1½"   1⅝"

FIGURE 2-33 Finish nailer *(courtesy Senco Products, Inc.)*

FIGURE 2-34 Pin nailer or brad nailer

FIGURE 2-35 Biscuit joiner

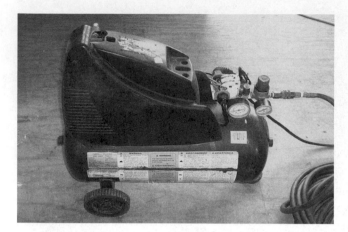

FIGURE 2-36   Air compressor

gauged in pounds per square inch (**psi**). Pressure output can be adjusted on air compressors and should be set to comply with the manufacturer's recommendations for the particular tool being used. Using too much pressure can cause damage to the tool and is unsafe for the operator of the tool.

Air hoses connect to the air compressor, usually with a quick-connect coupling, to supply nailers and other tools with air. Moisture accumulates inside the tank of air compressors because of the compression of air; therefore, air tanks should be emptied daily, preferably after each use. Valves on the bottom portion of the tank are opened to release accumulated water. The valves are designed to turn using only hand pressure; using a wrench on a valve will ruin it.

MAINTAINING TOOLS

Care of tools is a sign of professionalism. Experienced finish carpenters know that proper care and maintenance of tools is essential for safety and efficiency reasons and is beneficial to their careers. Care and maintenance of tools will result in

- Longer tool life
- Safer operation
- Producing work of higher quality
- Increased productivity due to less down time

The following safety tips are things that should be done regularly to ensure that tools are safe and in proper working order.

 SAFETY TIPS

- Replace dull blades.
- Make sure safety features are working properly. Do not use a tool with a malfunctioning safety device.
- Inspect cords; make sure they are not frayed or worn.
- Oil tools according to manufacturers' suggestions.
- Keep tools clean and do not allow a buildup of sawdust to accumulate; such a buildup can affect safety features.

 # SUMMARY

- Tools can be broken down into two basic categories: hand tools and power tools.
- A tool that does not require electricity is a hand tool; one that requires electricity is a power tool.
- Proper care and maintenance of tools and equipment is essential for both safety and efficiency reasons.
- Safety features on tools should be checked on a regular basis to ensure they are working properly.
- Unplug tools before changing saw blades or performing any other type of work (e.g., checking safety features).

- Tools should be used only for their intended purposes.
- Safety glasses should be worn when operating any type of equipment.
- The cutting action of most tools is in a direction that is away from the body.
- Use a backing board to prevent smaller pieces of wood from being launched from a saw.
- Use push sticks to keep hands away from blades and cutter heads.
- All nails and screws should be countersunk.

KEY TERMS

chatter Ripples in the wood that are left behind when a router bit vibrates against the wood. Chatter is caused by using a dull router bit or overworking the router by moving it faster than what the motor can handle.

cutting an inch (also referred to as *burning an inch*) To measure by cutting an inch, begin the measurement directly on the 1-inch mark, and mark the desired measurement. Do not forget to add the cut inch to the overall measurement (e.g., instead of marking 12 inches, you would place the mark on 13 inches to make up for the inch that was cut).

hand tools Tools that are not powered by electricity, air, gas, or batteries.

level Flat, perfectly horizontal, in relation to gravity. A level is also an instrument used for checking things to see if they are level or plumb.

pilot hole A drilled hole that provides a path for a screw, bolt, or nail; a pilot hole prevents screws and nails from wandering and keeps wood from splitting.

plumb Perpendicular, straight up and down, at a right angle to something that is level.

power tools Tools that require electricity or some alternate source of power (e.g., battery, air, gas).

predrilling Drilling a hole, referred to as a **pilot hole**, prior to the insertion of a nail or screw. Predrilling provides a path for nails or screws to follow and prevents the wood from splitting.

psi Pounds per square inch. Air compressors supply pneumatic nailers and other tools with compressed air that is gauged in psi. Pressure output can be adjusted on air compressors and should be set to comply with the manufacturer's recommendations for the particular tool being used.

rip fence The adjustable guide on a table saw or router table. The rip fence is adjustable so that material can be ripped to an exact width.

shoe The base of a circular saw. It is adjustable so that materials of differing thickness can be cut.

REVIEW QUESTIONS

1. List five tools commonly used by finish carpenters.
2. Is it okay to bypass the safety features on certain tools?
3. Into what two categories can tools be classified?
4. A tool that requires electricity is referred to as what type of tool?
5. Name two results of proper care and maintenance of tools.
6. What category of tools does a block plane fall into?
7. What is the primary function of a thickness planer?
8. What should be done before changing a saw blade or performing any type of work on a tool or machine?
9. At what depth should table saw blades and skill saw blades be set?
10. What is chatter?
11. What can be done to eliminate chatter when using a router?
12. Name a tool, other than a spirit level, that could be used for marking level lines around a room.
13. What tool is used to cut decorative profiles in wood, especially along the edges of wood?
14. Why should nails and screws be countersunk?
15. What tool, in combination with a hammer, is used to countersink nails?
16. What is something that might interfere with a finish nail, causing it to curl out of the wood?
17. What tool is used for making irregular and radial cuts?
18. What category of tools does a coping saw fall into?
19. What is a scratch awl used for?

CHAPTER 3

Interior Doors and Casings

OBJECTIVES

After studying this chapter, you should be able to

- Distinguish between right- and left-hand doors
- Properly install doorjambs
- Trim around doorjambs
- Make a cased opening
- Cut hinge gains
- Hang doors on jambs
- Adjust doors
- Install doorjambs and door locks

INTRODUCTION

Without proper installation, doors will not function properly. This chapter discusses techniques for installing and trimming doors that will make them both attractive and functional. **Figure 3-1** illustrates the basic parts of a door. The finish carpenter must be able to properly install **doorjambs**, trim the doorjambs, cut hinge gains, hang doors, and build cased openings, all of which require a high level of skill. The door is housed within the doorjambs. Jambs are secured within the rough opening using shims and finish nails. **Shims** are wedge-shaped pieces of wood or plastic; sometimes several shims must be stacked together in order to fill the space between the jamb and rough framing. Doors are shimmed at several key areas (see the section *Installing Shims* later in this chapter). Hinges attach to both the door and the jamb. The **doorstop** is nailed onto the jambs and is designed to stop the door when it is being closed. **Casing** is molding attached to the doorjambs. Casing enhances a door's appearance and conceals the gap between the doorjamb and rough framing. **Hinge gains** are cut into both the door edge and the doorjambs. Hinge gains are cut to a depth equal to that of the hinge thickness, allowing the hinges to sit flush, or flat, with door and jamb surfaces. **Flush** is a term used to describe two surfaces that are even with one another (having no offset between two surfaces). A **cased opening** is an opening between rooms consisting of jambs and casing only. A cased opening does not have a door or doorstop. ■

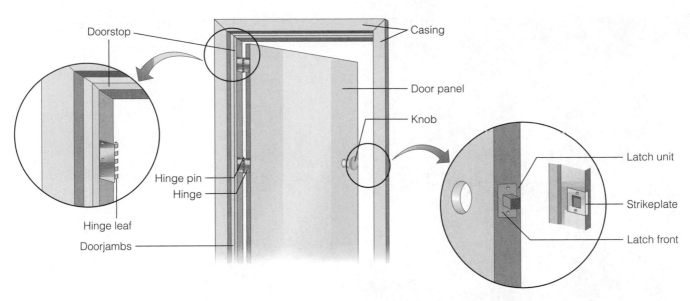

FIGURE 3-1   Basic parts of a door

TYPES OF INTERIOR DOORS

Some basic types of interior doors are single, double, bifold, pocket, and bypass doors.

Functional Types

Function describes how the door operates. A **single door** (**Figure 3-2**) consists of one door. A **double door** (**Figure 3-3**) consists of two doors hinged within a single jamb unit. A **bifold door** (**Figure 3-4**) can be either a single or double door. The bifold door is hinged in the center and slides open on a track versus swinging outward like a typical door. **Pocket doors** (**Figure 3-5**) operate on a track and can slide back into a cavity in the wall. They are good for areas of limited space. The pocket door is installed during the rough framing stages of building so that the cavity can be covered with wallboard. **Bypass doors** (**Figure 3-6**) operate on tracks and rollers as well. Each door runs on its own track; therefore, the doors are able to pass in front of or behind one another, opening within a limited space.

FIGURE 3-2   Single door

Double Doors, Bifold Doors, and Bypass Doors

These types of doors are usually wider and have a longer head jamb than single doors; however, they are installed using the same methods. The jambs of these doors must be plumbed and shimmed the same as those of a single door. The main difference between a single and double door (including bifold and bypass

FIGURE 3-3   Double door

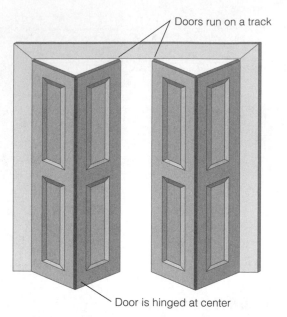

FIGURE 3-4   Bifold door

doors) is the addition of a center margin, or gap, between the two parts of a double door. The **margin** is the gap, or space, between the door and jamb or other door. Double doors also must receive an additional

shim in the center of the head jamb, between the jamb and the rough framing, to prevent any possible movement caused by the weight of the doors on tracks; and they have latching systems located at the center of the head jamb.

Pocket Doors

Pocket doors are normally installed during the framing stage of construction. This is because they are covered with sheetrock and finished with the walls.

FIGURE 3-5   Pocket doors

Top View

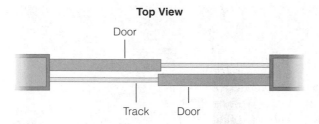

FIGURE 3-6   Bypass door

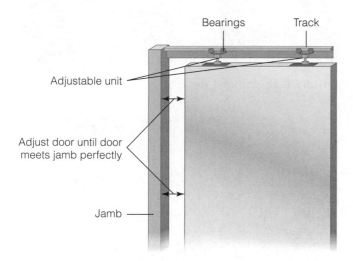

FIGURE 3-7   Adjusting pocket door

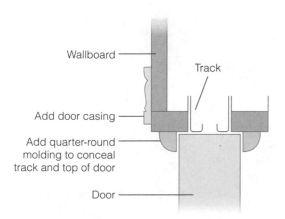

FIGURE 3-8   Completing trim on a pocket door

Trim carpenters attach roller hardware to the top of the door and hang the door, making proper adjustments (**Figure 3-7**). The doorstop and trim are added to complete the pocket door (**Figure 3-8**). Floor guides must be attached so that the door functions properly (**Figure 3-9**).

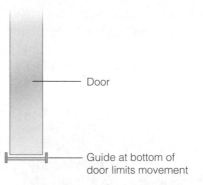

FIGURE 3-9   Floor guides on a pocket door

FIGURE 3-10   Smooth (flush) surface

Design Types

Design describes the appearance or texture of the door. Architectural design dictates the design of the door being used. The design type does not affect the method of installation. Some common design types of doors are smooth or flush (**Figure 3-10**), raised panel (**Figure 3-11**), louver (**Figure 3-12**), and French or glass (**Figure 3-13**).

Door Hand: Right- and Left-Handed Doors

Door hand, sometimes referred to as *door swing,* describes the direction in which the door opens. A simple way to tell whether a door is left- or right-handed is to stand with your back to the hinges; if the door swings to the right, it is a *right-hand* door, and if it swings to the left, it is a *left-hand* door (**Figure 3-14**).

> **NOTE**
>
>   Doors can also be classified as left-hand reverse and right-hand reverse. Doors normally open into the room (or inward). Reverse doors open outward. Physically, an LH (left-hand) door and an RHR (right-hand reverse) door are identical. The lock or the door hardware is what makes the difference. In a like manner, an RH (right-hand) door and an LHR (left-hand reverse) door are the same. ■

FIGURE 3-12   Louver design

FIGURE 3-11   Raised panel design

FIGURE 3-13   French (glass) door

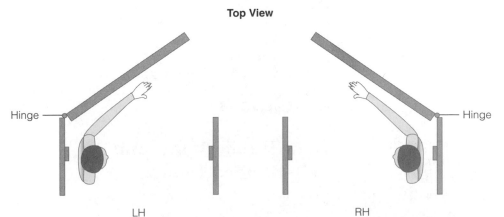

Top View

FIGURE 3-14   Door swing

Hinge

Hinge

LH

RH

Prehung Doors

Prehung doors are widely used in private and commercial industry. They come from the factory complete with hinges and predrilled holes for accepting the lock set. In some cases, the doors come with trim already attached to one side. Builders use this type of door for the simple reason that it saves time. A prehung door unit simply has to be set in the proper opening, plumbed, and nailed. Once nailed into position, shims are set in place and nailed, and the other side of the door is then trimmed.

Cased Openings

A cased opening connects two rooms. It consists of a trimmed jamb without a door. When building and installing a cased opening, the carpenter should assume that one day it may be equipped with a door. The unit needs to be plumb, square, shimmed, and sized to accommodate a standard-sized door (the largest size that the rough opening will allow you to make). (See the section on door sizes later in this chapter for information about standard sizes for doors.)

MATERIALS

Interior doors are constructed using various materials, both natural and manmade. Solid wood doors are usually used when a stain finish is desired. Solid wood doors are heavy and restrict noise flow between rooms.

Solid core describes a door that is solid all the way through. These doors are heavy and restrict noise flow. Solid core doors have a higher fire rating than hollow core doors (meaning that it would take longer for a fire to burn through a solid core door). **Hollow core** describes a door that is hollow. Solid core and hollow core doors are available in many different styles (e.g., raised, panel, flush, etc.). They are comprised of a compressed fibrous material that is similar to MDF. These doors usually come with a primer finish that is ready for accepting paint. They are attractive and cost-efficient, and hence are commonly used in new custom homes.

Door Sizes

The following list shows the most commonly used door sizes that are available:

Widths	Heights
1'-6"	6'-8"
2'-0"	7'-0"
2'-4"	
2'-6"	
2'-8"	
3'-0"	

The most commonly used sizes for interior doors are 2 feet 0 inches, 2 feet 6 inches, 2 feet 8 inches, and 3 feet 0 inches. The standard door height is 6 feet 8 inches.

Doorjambs

Doorjambs house the door itself. Hinges, casing, and doorstops are attached to the jambs. Both custom-made and prefabricated jambs should be installed carefully to ensure that the unit functions properly. Jamb material is typically ¾-inch thick. The width depends on the thickness of the wall.

Split Jambs

Split jambs are usually prefabricated and adjustable to fit varying wall thicknesses. The joint of the jamb is hidden by the doorstop (**Figure 3-15**). Split jambs are sometimes used between rooms where a transition in wood species occurs. One room may be finished with mahogany while the adjoining room is finished in oak. The split jamb is made to match each room. Setscrews are used to adjust the jambs to the wall thickness. Split jambs are installed in the same manner as prehung door units.

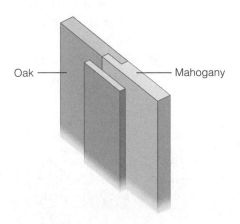

Oak — Mahogany

FIGURE 3-15 Split jamb

PROCEDURES: INSTALL INTERIOR DOORS

 SAFETY TIPS

- Wear OSHA-approved safety glasses, footwear, and hard hat.
- Use only GFCI receptacles, use proper gauge extension cords, and never work around wet areas.
- Keep blades sharp, and keep equipment and workspace clean and free of safety hazards.
- Pay strict attention when operating any tool, and always keep safety features intact and functioning properly.

3-1 INSTALLING DOORS AND CASED OPENINGS

 **NOTE**

A level is absolutely necessary when installing doors.

FIGURE 3-16   Measuring for jamb width

The first step in installing either a door or a cased opening is to measure the width needed for the jamb stock. A wall constructed with 2 × 4 studs measures *roughly* 4½ inches when covered with ½-inch wallboard on either side. A wall constructed with 2 × 6 studs measures *roughly* 6½ inches when covered with ½-inch wallboard on either side (**Figure 3-16**). Rip jamb stock on the table saw at approximately ⅛ inch over the size needed so that saw marks can be planed or sanded off, leaving a smooth edge.

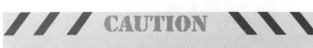 **CAUTION**

Always adjust the depth of saw blades prior to turning on the machine. Set the blade cutting depth according to the thickness of the material being cut.

Door Trim and Casing

Door trim should sit flush on finished floors. Use intersecting reveal marks to get a measurement for the head casing. **Reveal** is the amount of offset between two components. The marks made for the reveal will determine how far away the edge of the casing will be from the edge of the jamb. Reveal marks can be made using a tape measure, combination square, or gauge block. **Gauge blocks** can be cut from scrap material and are used when there is a need to make many marks of the same dimension. Using a gauge block is efficient in that it saves you from having to repeatedly use the tape measure. Use glue on miter joints and cross-nail where the head casing meets the side casing (**Figure 3-17**). Avoid erratic nailing patterns when fastening the casing (**Figure 3-18**). Uniform nailing patterns reflect professionalism and make the painting contractor's job easier.

(a)

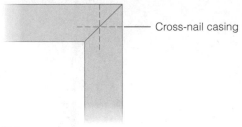

Cross-nail casing

(b)

FIGURE 3-17 Use wood glue and cross-nail miter joints.

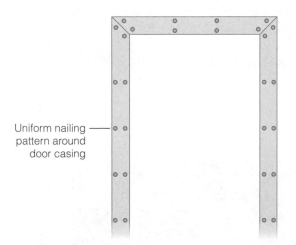

Uniform nailing pattern around door casing

FIGURE 3-18 Avoid using erratic nailing patterns on door casing.

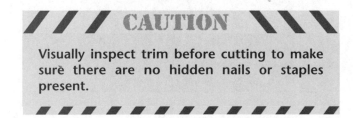

///// CAUTION \\\\\

Visually inspect trim before cutting to make sure there are no hidden nails or staples present.

/////   \\\\\

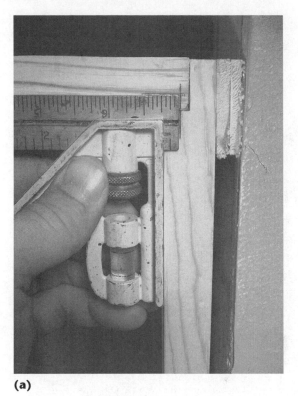

(a)

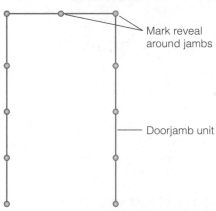

Mark reveal around jambs

Doorjamb unit

(b)

FIGURE 3-19 Use a gauge block or combination square to mark a consistent reveal around doorjambs at several points.

Following are the steps for applying door casing:

1. Using a gauge block or combination square, make light reveal marks between ⅛ inch and ¼ inch) around the doorjambs (**Figure 3-19**).

2. Measure and cut the headpiece and install it (**Figure 3-20**).

3. Cut the sidepieces to match the headpiece height (**Figure 3-21**).

4. Fasten the casing with finish nails, using the reveal marks as a reference.

5. Sand off reveal marks.

FIGURE 3-20 Measure, cut, and install head casing.

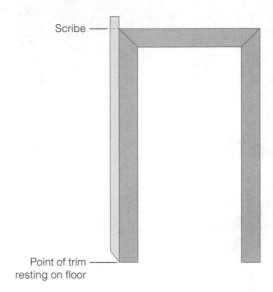

FIGURE 3-21 Cut side casing to fit.

Rosettes and Plinth Blocks

Doors are sometimes trimmed with rosettes and plinth blocks. **Rosettes** are decorative square blocks positioned where the head casing meets the side casing. The rosette acts as the capital of a column and is centered on the casing. Casing meets rosettes with a square cut (**Figure 3-22**). Rosettes are fastened with finish nails.

Plinth blocks are positioned beneath the side casings (**Figure 3-23**) and provide a transitional point between the side casing and the base molding. Plinth blocks, depending on the thickness, may have to be predrilled and secured with screws (**Figure 3-24**). Afterward the hole receives a wood plug, which is sanded flush with the block's surface.

FIGURE 3-22 Casing meets rosettes with a square cut.

FIGURE 3-23 Plinth blocks are positioned beneath the side casing.

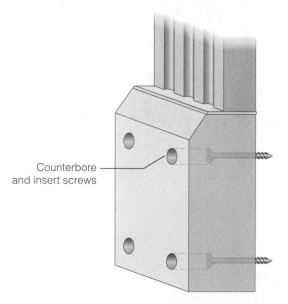

Counterbore and insert screws

FIGURE 3-24 Installation of plinth block (counterbore and secure with screws)

3-2 MARKING JAMBS FOR DOOR CASING

Jambs need to be marked so that trim is uniform when nailed on. To mark uniform lines around the jambs, use a tape measure, a gauge block, or a combination

square (refer to Figure 3-19). The door casing is offset and is not flush with the edge of the jamb. The amount of offset is called the *reveal*. A reveal between ⅛ inch and ¼ inch is common on doorjambs. Mark where the head jamb and side jambs intersect and at spaced intervals around the unit. Casing will be fastened according to the reveal marks.

3-3 INSTALLING DOORJAMBS

Rough openings are made anywhere from 2 to 2½ inches larger than the door, which will leave enough room between the jamb and rough framing for shims and any adjustments that have to be made. Side jambs may rest on the floor; later a flooring contractor will cut them to the finished height so that tile or hardwood may be slid beneath them. If you are installing jambs on a floor that is already finished, both side jambs will need to sit fully on the finished floor.

Following are the steps for installing doorjambs:

1. Set jamb unit in opening (**Figure 3-25**).
2. Check head jamb for level (**Figure 3-26**).
3. If head jamb is not level, adjust the unit by either shimming up the low side or trimming the high side (**Figure 3-27**).
4. Shim the jambs on either side of the head jamb (see the section *Installing Shims* later in this chapter).
5. Nail through the shims once the head jamb is level, making sure the jamb is flush with the wallboard (**Figure 3-28**).

FIGURE 3-25 Set jamb unit in opening.

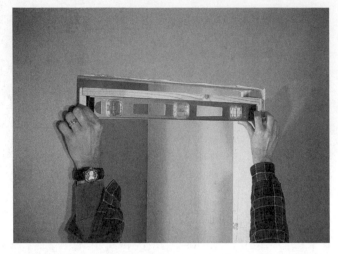

FIGURE 3-26   Check head jamb for level.

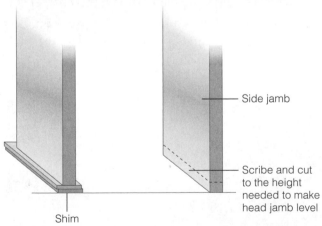

FIGURE 3-27   Adjust unit to make head jamb level.

6. Shim the side jambs and check for plumb, nailing through the jambs at shim locations when plumb (**Figure 3-29**).

7. Scribe jambs in preparation for door casing (see the earlier section *Marking Jambs for Door Casing*).

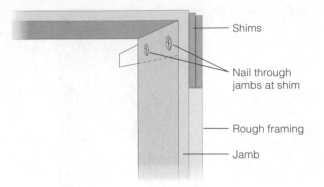

FIGURE 3-28   Make sure jambs are flush with wallboard, and then nail through jamb at shim location.

(a)

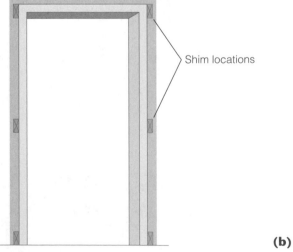

(b)

FIGURE 3-29   Shim and nail jambs at several locations when plumb and straight.

Installing Side Jambs and Head Jambs

Cut the side jambs and head jambs to the length needed. Side jambs are typically 81¾ inches on a standard door. Side jambs are **rabbetted** to create a shelf for the head jamb to rest on. The head jamb length depends on the size of the door being used. When cutting the head jamb, allow extra for the rabbetted shelf and gap, or margin, between the door and side jambs. To get the length needed for the head jamb, use the following equation:

(width of door) + (total width of rabbets)
   + (space between door and jamb)
   = length of head jamb

$36" + \frac{3}{4}" + \frac{3}{16}" = 36\frac{15}{16}"$ (length of head jamb)

Rabbetting Side Jambs

Side jambs can be rabbetted using a router, or they can be rabbetted with a circular saw and chisel. When you are using a router, use a ¾-inch straight fluting bit and set it to a depth of ⅜ inch (which is exactly half the thickness of the jamb stock). Following are the steps to follow in rabbetting side jambs with a router (**Figure 3-30**):

1. Mark a square line ¾ inch from the end of the side jamb.
2. Clamp a straight edge to guide the router.
3. Set router depth to ⅜ inch. (Test the depth on a piece of scrap wood.)

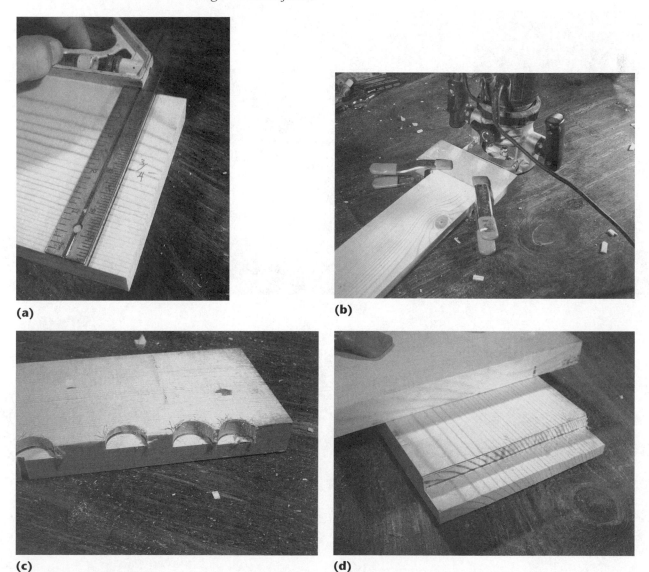

(a)   **(b)**   **(c)**   **(d)**

FIGURE 3-30 Rabbeting side jamb with a router: (a) Mark a square line 3/4" from the end of the side jamb. (b) Clamp a straight edge to guide the router. (c) Remove excess wood from side jamb, creating the rabbetted shelf. (d) Finished rabbet.

CAUTION

Clamp materials firmly using spring or wood clamps. Always hold router with both hands.

4. Remove excess wood from side jamb, creating the rabbetted shelf.

5. Finished rabbet.

Following are the steps to follow in rabbetting side jambs with a portable circular saw (**Figure 3-31**):

1. Mark a square line ¾ inch from the end of the side jamb.

2. Set saw to ⅜ inch. (Test setting on a piece of scrap wood.)

3. Make a series of cuts. (Be sure to leave a pencil line.)

4. Remove all wood with a chisel.

5. Finished rabbet.

Once the side jambs have been rabbetted, the head jamb can be cut to length and nailed in place. Apply wood glue to the shelf area before nailing on the head jamb (**Figure 3-32**). Do not use jambs that are

(a)

(b)

(c)

(d)

FIGURE 3-31   Rabbeting side jamb with a circular saw and chisel: (a) Mark a square line 3/4" from the end of the side jamb. (b) Make a series of cuts. Be sure to leave a pencil line. (c) Remove all wood with a chisel. (d) Finished rabbet.

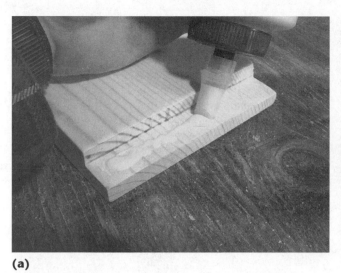

(a)

(b)

FIGURE 3-32 Apply wood glue to shelf area before attaching head jamb.

bowed or checked. Damaged jambs should not be used. Exposed edges of jambs should be lightly sanded with fine-grit sandpaper (100 to 220 grit). Slightly rounded edges accept paint better and are less likely to become damaged.

3-4 SPACING DOORS AND JAMBS

The spacing between the door and the jamb, sometimes referred to as the margin, should be no less than ⅛ inch on the latch side and ¹⁄₁₆ inch on the hinge side. The margin at the top should be ⅛ inch. Door bottoms are usually ⅝ inch above the finished floor (**Figure 3-33**). Margins allow for expansion and contraction due to seasonal changes. Inadequate margins will result in a door that rubs or sticks against the jamb. Margins should be *even* and *consistent* around doors.

Door edges should be beveled so that there is constant clearance between the door and the jamb while the door is being opened and closed (**Figure 3-34**). A bevel ensures that the door will not make contact or rub while being opened or closed. The widest point of the bevel should be on the side of the door making contact with the doorstop. Doors are beveled anywhere from ¹⁄₁₆ inch to ⅛ inch.

FIGURE 3-33 Margins around a door

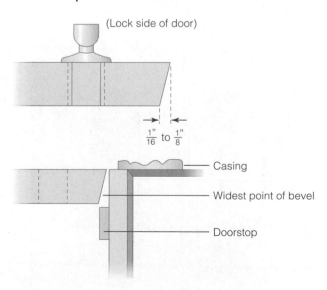

FIGURE 3-34 Beveled edge on a door

3-5 INSTALLING PREHUNG DOORS

Prehung doors are commonly used in today's building industry because they are efficient. Prehung doors are fully complete, meaning that the jambs have been assembled, hinge gains have been cut and both the door and jamb have been fitted with hinges, holes have been bored for the latch bolt and doorknob, a mortise has been cut into the jamb for receiving the strike plate, doorstop has been attached, and, in most cases, the doorjambs have been fitted with casing, which is already attached to one side (the casing for the other side usually comes with the door; it is attached on the side of the doorjamb and has to be removed before installation of the door). A prehung door simply has to be installed in the proper opening, which saves the finish carpenter a lot of time. Before installation, refer to plans or blueprints to see if the opening receives a left- or right-hand door.

 NOTE

There is usually a member referred to as a **spreader** at the bottom of prehung doors, which helps support the door during shipping. Before installation, the spreader must be removed and any staples or other fasteners that are left in the bottom of the jamb should be pulled out (not driven in). A flooring contractor may eventually have to cut the bottom of the jambs so that the flooring will fit beneath both the jambs and the casing (fasteners that are left in the bottom of the jamb become a safety hazard and can possibly damage the equipment of the flooring contractors). ■

Figure 3-35 shows the steps of installing a prehung door.

1. Plumb hinge side and nail hinge-side casing when door is plumb.

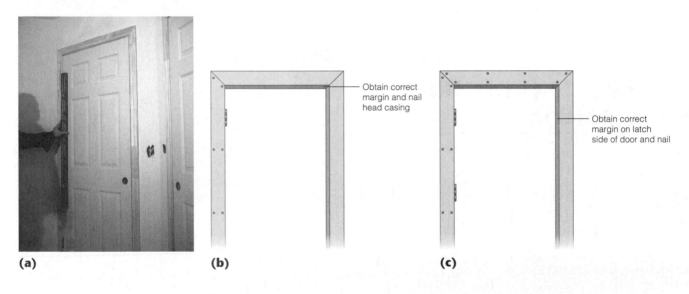

(a)   **(b)**   Obtain correct margin and nail head casing   **(c)**   Obtain correct margin on latch side of door and nail

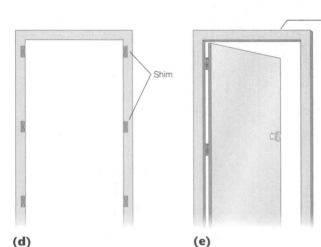

(d)   Shim   **(e)**   Attach casing to the other side of door

FIGURE 3-35 Steps in installing a prehung door: (a) Plumb hinge side of door; (b) Obtain correct margin above door and nail head casing; (c) Obtain correct margin on latch side of door; (d) Shim door; (e) Install casing.

2. Obtain correct margin above door and nail head casing.

3. Obtain correct margin on latch side of door, nailing casing when good.

4. Shim door jamb.

5. Install casing to remaining side.

/// / **CAUTION** \ \\\

Finish nails can curl out of wood due to knots and grain pattern. Keep hands and fingers a safe distance away when using finish nailers. Always wear safety glasses.

///////////////

3-6 INSTALLING SHIMS

All doors, including prehung doors, should be shimmed inside the rough opening. Without shims, over time the door will cease to function properly as the weight of the door pulls against the hinges and jambs. The result of not using shims will be a door that sticks or rubs against the opening and a door latch that will eventually quit working properly.

Shims of plastic or wood are wedge-shaped and should be stacked in such a way that they fill the space between the jamb and rough opening evenly (**Figure 3-36**). Shim doors at key areas, as shown

earlier in Figure 3-29. Double doors and doors with longer head jambs should receive an additional shim at the center of the head jamb where latches and tracks have a tendency to cause stress and movement (**Figure 3-37**). Protruding portions of the shim should be cut off flush with the wallboard so that they do not interfere with the door casing. Cut excess portions of shims off using a utility knife or sharp chisel. Once the door is secure and shims are in place, remove the center hinge screw in each hinge and replace it with a 3-inch brass screw to further secure the unit (**Figure 3-38**).

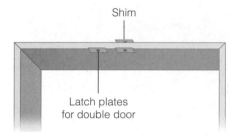

FIGURE 3-37 Shim the center of longer head jambs, as on a double door.

FIGURE 3-36 Stack shims to fill the space evenly between the jamb and rough opening.

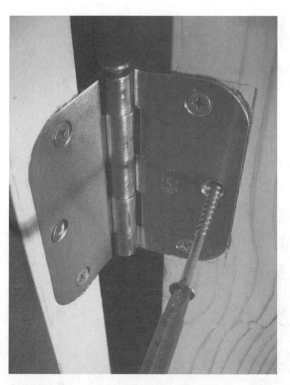

FIGURE 3-38 Inserting a 3-inch brass screw through the hinge to secure the unit

3-7 CUTTING HINGE GAINS

Matching hinge gains are routed into doors and jambs. Gains are cut with either a router or chisel. Gains allow the hinge to sit *flush* with the door or jamb surface (**Figure 3-39**). A door and jamb template is usually used to cut matching gains. The template ensures accuracy of matching gains on both door and jamb. Prehung doors are already equipped with hinge gains

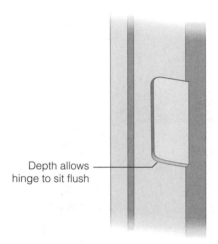

Depth allows hinge to sit flush

FIGURE 3-39   Hinge gains (cutting and fitting hinge)

FIGURE 3-40   Drill pilot hole and then insert hinge screw

and hinges. Brass screws secure the hinges to the door and to the jambs. A door is hung once both the door and the jamb have been equipped with hinges and the hinges have been joined together and secured with a hinge pin.

Once hinge gains have been cut with a router or chisel, drill a pilot hole to keep hinge screws centered and to prevent the jamb from splitting (**Figure 3-40**). The hinges are joined together by a hinge pin slid in from the top (**Figure 3-41**). Make sure that the gains cut on the door and the jamb match and that the hinge is flush with the surface.

Hinge locations are shown in **Figure 3-42**. The top hinge is centered 9 inches from the top, the bottom hinge is centered 12 inches up from the bottom, and the middle hinge is centered between the top and bottom hinges. The depth of the hinge should be set so that the door can be opened fully (**Figure 3-43**).

If a door and jamb template is not available, cutting gains into doorjambs can be approached with either of the following two methods.

Method One

Figure 3-44 illustrates the following steps:

1. After rabbetting a shelf into the side jamb, measure and mark hinge locations as shown in Figure 3-42, lightly scribing layouts onto jambs.

FIGURE 3-41   Hinge pin being inserted

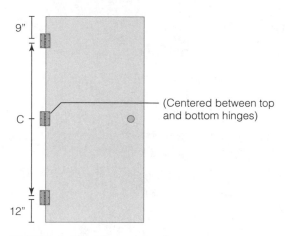

9"

C (Centered between top and bottom hinges)

12"

FIGURE 3-42 Hinge locations

FIGURE 3-43 Set hinges so that doors can open fully.

2. Use a combination square to maintain consistent depth in square layouts at all hinge locations.

3. Once the hinge layouts have been scribed, use the hinges themselves to trace the corners of the hinges (which are usually round, although some hinges are square).

4. Use a router with a straight fluting bit, set to a depth equal to the thickness of the hinge, to cut out all hinge gain locations. (Test the depth setting on a piece of scrap material prior to making the final cuts.)

5. If using hinges with square corners, cut out the remainder of the hinge gains with a sharp wood chisel.

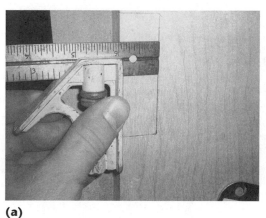

(a)

(b)

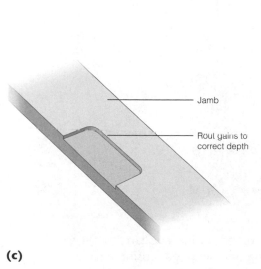

Jamb

Rout gains to correct depth

(c)

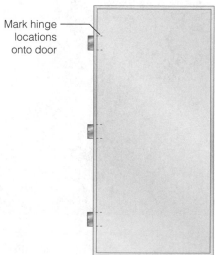

Mark hinge locations onto door

(d)

FIGURE 3-44 Method of routing hinge gains: (a) Use combination square to lay out gain; (b) Use hinge to trace round corner; (c) Hinge gain; (d) Corresponding gain locations marked onto door.

6. Assemble the jambs, and then install the jambs completely. (See the section earlier in this chapter on installing doorjambs.)

7. To mark corresponding hinge-gain locations onto the door, place the door in the opening and scribe the hinge gain locations onto the door. (Remember to shim the door up until there is a ⅛-inch margin between the top of the door and the jamb.)

8. Check to make sure the combination square is still set to the proper depth, and then use it to finish scribing the layouts onto the door.

9. After cutting the layouts on the door, drill pilot holes and screw the hinges into place on both the door and the jambs.

10. Attach doorstop.

11. Then hang the door, joining the hinges by sliding hinge pins in from the top.

Method Two

Figure 3-45 illustrates the following steps:

1. Assemble the jambs and install them, making sure you plumb and shim the jambs. (See the section *Installing Doorjambs* earlier in this chapter.)

2. Set the door within the jambs, making sure the door is shimmed up to a height where there is a ⅛-inch margin between the top of the door and the jamb.

3. Mark hinge gain locations onto both the jamb and door at the same time.

4. Use the combination square to lay out the hinge gains on the door and the jamb.

5. Rout hinge gains with a router set to the proper depth.

6. Install hinges.

7. Attach doorstop.

8. Mount door, securing hinges with the hinge pins.

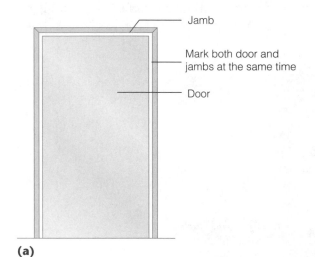

(a)

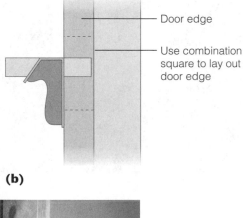

(b)

(c)

FIGURE 3-45 Method of routing hinge gain: (a) Mark hinge locations on both door and jamb; (b) Use combination square to lay out hinge gains; (c) Free-handing the cutting of a hinge gain using a router.

3-8 ADJUSTING DOORS

Doors need adjustments when there is not a consistent margin between the door and the jamb. Slight adjustments are sometimes enough to correct the problem by creating more or less margin as needed. To achieve a consistent and uniform margin (space) between the door and the jamb, use one of the following methods to adjust the door:

- Remove shim and tighten the 3-inch screw in hinge.

- Rout the hinge gains deeper.

- Plane the door edges (**Figure 3-46**).

- Insert cardboard behind the hinges and retighten the hinge screws.

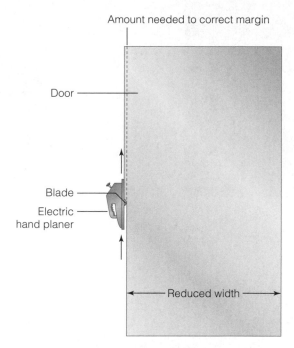

FIGURE 3-46   Plane door edges

> **NOTE**
>
> Either of these methods can be used to cut precision hinge gains if one does not own a hinge-gain routing template. Method number one is slightly easier because the gains on the jambs can be cut on the worktable. Method number two requires the gains to be cut with the jamb sitting in a vertical position, making it more difficult to use the router and to see the cutting path of the bit. Method number two is something that would have to be done in a remodeling situation, when adding a door to an existing cased opening. ■

DOORSTOPS

Doorstops are trim that is nailed around the inside portion of the doorjambs to provide a stopping point for doors. Doorstops are usually rounded slightly on one side and square on the other. The square side makes contact with the door. On the hinge side of the door, the stop has about a $1/32$-inch clearance between it and the door so that the door does not rub against it while being opened and closed. The stop on the lock side makes contact with the door. The headpiece of the stop needs to connect perfectly with the side stops.

Molded doorstops (doorstops with profiles) should be joined with a miter cut. Square doorstop material can be either mitered or square-butted (**Figure 3-47**). Stops on prehung doors are joined with butt joints and sit flush with the finished floor or with the bottom of the jambs on an unfinished floor. **Figure 3-48** shows a doorstop that has been installed.

DOOR LOCKS

Figures 3-49 through **3–52** show some types of knobs and lock sets used on interior and exterior doors. Lock sets used include the mortise lock set, cylindrical lock set, tubular lock set, and unit lock set. If the door is not predrilled for a lock set, you will need to bore it with a hole saw (**Figure 3-53**). Hole saws are available in many different sizes.

For mortising for the faceplate, use a faceplate mortise maker. This tool makes a square cut that fits most strike plates. Tapping the tool makes a square cut, which is then chiseled out to a depth that will allow the strike plate to sit flush with the door edge. A chisel can be used for the whole process of cutting out for the faceplate (**Figure 3-54**). Use the faceplate as a template for marking the door edge and then chisel out the layout to a depth that will allow the faceplate to sit flush with the surface of the door edge (**Figure 3-55**).

(a)   (b)

FIGURE 3-47   Applying doorstop: (a) Square doorstop can be joined with a butt joint; (b) Molded doorstop should be mitered.

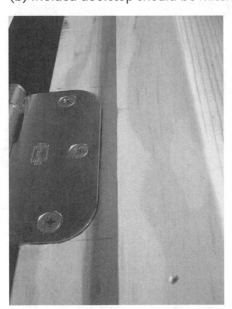

FIGURE 3-48   Doorstop after installation

FIGURE 3-50   Cylindrical lock set

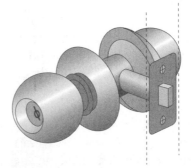

FIGURE 3-51   Tubular lock set

FIGURE 3-49   Mortise lock set

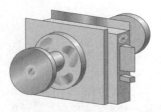

FIGURE 3-52   Unit lock set

(a)

(b)

FIGURE 3-53 Boring on both sides of door with a hole saw to receive lock set

FIGURE 3-54 Mortising for the faceplate using a chisel

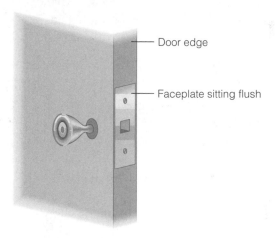

Door edge

Faceplate sitting flush

FIGURE 3-55 Faceplate sitting flush with door surface

 NOTE

Use a wedge to secure the door open while working on the door. ∎

Installing a Cylindrical Lock Set

The following are instructions for installing a cylindrical lock set:

1. Use the knob template to mark the door. (Use a square to lightly scribe lines at knob center.) Knobs are usually centered between 36 and 38 inches off the floor. Mark the center of the 2⅛-inch hole and the center of the ⅞-inch hole.

2. Drill the 2⅛-inch hole using a hole saw or boring bit (drill in from *both* sides to prevent exit damage from the boring bit). Drill the ⅞-inch hole into the door

edge using a hole saw or a spade bit. Chisel out the door edge so that the latch front sits flush with the surface of the door edge. Install the latch unit.

3. Install the strike plate, mortising the doorjamb at the proper height. (The strike plate height should be centered the same as the height of the door latch.) Use the strike plate as a template, tracing its outline with a scratch awl or a sharp pencil. Mortise the area to a depth that will allow the strike plate to sit flush with the jamb surface. Set the strike plate in position, mark the latch bolt area, and mortise the area to an adequate depth (one that will allow the latch bolt to extend fully).

4. Install the lock assembly and doorknobs, using the screws that are provided with the unit. If using a keyed latch, make sure the key side is installed

on the correct side of the door, as some lock sets are made differently for right- and left-hand doors. The name brand of lock sets should appear with the right side up.

THRESHOLDS

Generally, interior doors are not required to have thresholds. Thresholds are designed for exterior doors. Their purpose is to keep out weather and other outside elements, such as insects and dust. **Figures 3-56** and **3-57** show two different types of thresholds. Some thresholds are adjustable, which allows the proper height to be set by the installer. Height is adjusted by turning the setscrews. Other doors have thresholds that seal themselves when contact is made between the door and an expandable rubber membrane. Automatic door bottoms are available for use on interior doors. They reduce noise and

air flow. The automatic door bottom is movable, making contact with the floor when the door is closed and rising when the door is opened.

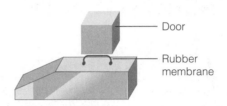

FIGURE 3-56 Adjustable threshold

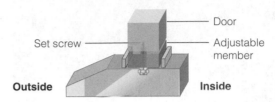

FIGURE 3-57 Self-sealing threshold

SUMMARY

- The appearance of a door does not affect its method of installation.
- A level is absolutely necessary when installing doors.
- Keeping a margin of no less than ⅛ inch between the door and jambs allows for the swelling and contracting of wood due to seasonal changes, doing away with the possibility of doors sticking.

- Erratic nailing patterns make the painting contractor's job more difficult.
- When setting a door, besides keeping it plumb, take special care to ensure that the margins between the door and the jamb are even and equally spaced.
- One way to ensure that a door will remain functional is to place shims at key locations. Doors without shims may cease to function properly.

KEY TERMS

bifold door Can be either a single or double door. The bifold door is hinged in the center and slides open on a track versus swinging outward like a typical door.

bypass door Operates using tracks and rollers. Each door runs on its own track; therefore, the doors are able to pass in front of or behind one another, opening within a limited space.

cased opening An opening between rooms consisting of jambs and casing only. A cased opening does not have a door or doorstop.

casing Molding attached to door and window jambs.

door hand The direction in which a door opens. Sometimes referred to as *door swing*.

doorjamb A jamb that houses the door, surrounding it on three sides. Hinges are screwed to the jamb as well as the door. Jamb stock is usually ¾ inches thick.

doorstop Is nailed to the doorjambs and provides a stopping point for the door.

double door Two doors hinged within a single jamb unit.

flush Two surfaces that are even with one another, having no offset between the two.

gauge blocks Blocks that can be cut from scrap material and are used when there is a need to make many marks of the same dimension. Using a gauge block is efficient in that it saves one from having to repeatedly use the tape measure.

hinge gains Sometimes referred to as *hinge mortises*. Hinge gains are cut into both the door edge and doorjambs. Hinge gains are cut to a depth equal to that of the hinge thickness, allowing hinges to sit flush, or flat, with door and jamb surfaces.

hollow core A door that is hollow.

margin The gap, or space, between the door and jamb or other door.

plinth blocks Blocks that is positioned beneath the side casings. The plinth block provides a transitional point between the side casing and base molding.

pocket doors Doors that operate on a track and can slide back into a cavity in the wall. They are good for areas of limited space.

prehung doors Doors that come complete and are ready to install.

rabbet A shelf-type cut at the end or along the edge of a board. Head jambs join side jambs with a rabbet joint.

reveal The amount of offset between two components.

rosettes Decorative blocks installed where side casings meet head casings. Rosettes are used on both windows and doors.

shims Wedge-shaped pieces of wood or plastic. Used for shimming doors and windows.

single door One door.

solid core A door that is solid all the way through. A solid core door reduces noise flow and has a higher fire rating than a hollow core door.

spreader Supports the door and protects it during shipping. Spreaders must be removed before installation.

REVIEW QUESTIONS

1. What areas require shimming on doors?
2. What is an ideal margin for a door?
3. What is a door called that swings in neither direction, but rather opens by sliding back into a wall cavity?
4. Describe how you can tell the difference between a left-hand and a right-hand door.
5. Name three types of interior doors.
6. Why is centering a door within a rough opening important?
7. When plumbing a door, where is the level placed?
8. Briefly describe a cased opening.
9. What could possibly happen to a door that is not shimmed?
10. At what point during the installation process is a door shimmed?
11. Who decides if a door is in need of adjustment?
12. In order to adjust the door within the opening, the hinges can be shimmed using what?
13. How wide should the jamb stock for a door be ripped on the table saw?
14. Unlike a single door, a double door needs to be shimmed in one additional area. Where does the extra shim need to be placed?
15. What length screw can be placed through the hinge to adjust or secure a door?
16. What is another way to adjust a door besides placing a screw through the hinge?
17. When speaking of the appearance of a door, what is meant by the term *flush*?
18. What do exterior doors (unlike interior doors) have at the bottom that is designed to keep out weather and other outside elements?

CHAPTER

4 Base Molding

OBJECTIVES

After studying this chapter, you should be able to

- Prepare a jobsite for base molding
- Measure rooms for base molding
- Record measurements using numbers and symbols
- Make accurate 45- and 22½-degree cuts
- Cope inside corner joints and splice trim
- Stack bases to appear as one molding

INTRODUCTION

Base molding, sometimes referred to as baseboard, is installed after doors have been installed and trimmed with casing. Base is installed where floors and walls meet, covering the edges of flooring materials (**Figure 4-1**). Base butts into door casing. It is **mitered** at outside corners and **coped** at inside corners. A board is mitered when it is cut at any angle; an angled cut that allows two pieces of trim to join at a corner or splice. When a joint is coped, trim is cut in a way that allows it to fit precisely over an adjoining piece of molding; coping is cutting the trim's profile, allowing it to fit over an adjoining trim with the same profile. A coped joint, if done correctly, looks far better than a mitered joint.

When installing base molding, keep in mind that floors are not always level, walls can be out of plumb, and inside and outside corners are not always perfect. This chapter will cover basic situations concerning base molding, such as how to record measurements in a way that will reduce confusion. There can be hundreds of individual pieces of molding for a single home or jobsite that will have to be measured, cut, and installed properly. ■

FIGURE 4-1 Base is installed where floors and walls meet.

JOB SAFETY

As always, safety must be considered. The following safety tips apply to all cutting activities in finish carpentry work.

SAFETY TIPS

- Always cut the face side of the trim, letting the blade exit through the back side.
- Use a backing block when cutting small pieces of trim. The backing block will keep small pieces from becoming damaged and from being thrown from the saw. Pieces thrown from the saw can cause serious injury.

TYPES OF BASE MOLDING

Other trim can be incorporated with base molding to achieve a stacked trim. Shoe molding is sometimes installed along the bottom of the baseboard and rests firmly on top of the finished floor. Sometimes baseboard is installed on top of spacers or shims so that the finished flooring can be slid beneath the molding.

MATERIALS

As noted in Chapter 1, a wide variety of materials are used for base molding, such as stain-grade wood, paint-grade wood, and manufactured fiberboard materials. Interior molding is milled from both softwood and hardwood. Hardwood is used almost exclusively for stained finishes; however, both softwood and hardwood can be stained. Molding composed of manmade materials, such as MDF, is available and can be used when finishing with paint. Plastic and PVC moldings are also available.

PROCEDURES: INSTALL BASE MOLDING

SAFETY TIPS

- Always keep hands a safe distance away from the blade.
- Wear safety glasses when cutting trim.
- Cut in a direction that is away from the body.

See the section on tool safety in Chapter 2 for other safety tips to keep in mind when cutting materials.

4-1 PREPARING FOR BASE MOLDING

Make sure floors are clean and free of debris that could affect baseboard measurements and installation. An excessive buildup of joint compound in corners can cause measurements to be inaccurate and may not allow trim to be seated properly (**Figure 4-2**). Buildup in corners and on floors can be scraped away with a chisel or putty knife (do not damage wallboard or texture when scraping buildup out of corners).

Floors should be marked to indicate the location of all plumbing pipes and studs. Visually inspect to see that all plumbing locations have been marked. Finish nails can penetrate plumbing pipes, which can cause water leaks and costly damage. Protective metal plates are normally installed behind the wallboard to protect pipes. Since this is not always the case, you should be careful when nailing around plumbing locations (**Figure 4-3**).

FIGURE 4-2   An excessive buildup of joint compound in a corner

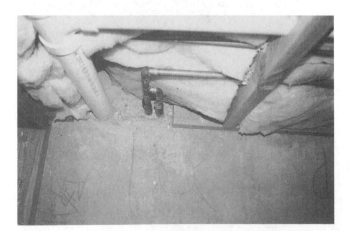

FIGURE 4-3   Floor should be marked to indicate plumbing locations.

4-2 MEASURING BASE MOLDING

The task of installing base molding begins with properly measuring the room to be trimmed. Using a method when measuring, cutting, and installing base reduces the confusion of dealing with a large number of individual pieces. Learning a method or a system that can be repeated from one room to the next results in simplicity.

The example in **Figure 4-4** shows the direction one can go in when measuring and installing base. Shown also are symbols used when recording measurements for molding (**Figure 4-5**). Working in the same direction in each room will reduce confusion because there will never be a question as to where the trim starts and where it ends. Also, it is necessary to continue

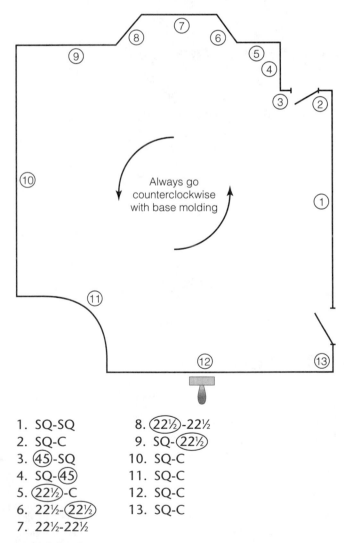

Always go counterclockwise with base molding

1. SQ-SQ
2. SQ-C
3. 45-SQ
4. SQ-45
5. 22½-C
6. 22½-22½
7. 22½-22½
8. 22½-22½
9. SQ-22½
10. SQ-C
11. SQ-C
12. SQ-C
13. SQ-C

FIGURE 4-4   Room and direction of installing base

Legend:	
SQ = Square	(22½)= Outside 22½
C = Cope	(45) = outside 45°
22½ = Inside 22½°	(+) or (−) = Add or
45 = Inside 45°	Subtract ⅟₃₂"

FIGURE 4-5 Numbers and symbols used for recording measurements

FIGURE 4-6 Number the back of each piece of trim as it is cut.

Measurements for M Bedroom
1. 10' SQ-SQ
2. 6" SQ-C
3. 4" (45)-SQ
4.

FIGURE 4-7 Recorded measurements for trim

working in one direction, since a coped piece of trim is nailed on top of the previous one. Numbering the back of each piece of trim as it is cut will help when it comes time to place the trim where it belongs (**Figure 4-6**). Working in a counterclockwise direction makes the act of *coping* easier (a left-handed person may want to work in a clockwise direction). See the section on coping trim that appears later in this chapter.

Record the measurement for each piece of trim (**Figure 4-7**). Use the symbols shown in Figure 4-5. The symbols will serve as reminders when cutting trim at the miter saw by showing which end of the trim receives what type of angle (**Figure 4-8**). Some people prefer to measure and cut one piece of trim at a time, and then install it, while others prefer to measure an entire room first, and then cut and install all the pieces. Some prefer to measure the entire house for base before cutting and installing any of it. Experience will help you determine which method works best for you in completing a job safely and efficiently.

When taking measurements, always hold the tape measure straight and do not permit it to sag. Sometimes while taking measurements for base it is helpful to use a block to support the tape (**Figure 4-9**). The block also holds the tape measure up off the floor, which keeps the tape from extending beneath the sheetrock (sheetrock is often held up ½ inch above the floor). Make sure inside and outside corners do not have an excessive buildup of joint compound, which can produce an inaccurate measurement and may prevent trim from seating properly.

Do not fold or bend the tape measure when taking inside measurements. Butt the end of the tape against the wall or surface, and butt the back of the tape housing against the other wall or surface; then read the measurement and add the length of the tape housing (**Figure 4-10**). The length of the tape housing is usually written on the bottom side of the tape housing. This is an extremely accurate method of taking inside measurements, much more accurate than folding or bending the tape into the corner.

Record measurements and symbols as discussed earlier to cut down on trips to the miter station, making efficient use of time. Work in one direction (clockwise or counterclockwise) to develop a system or method that eliminates the confusion of dealing with many pieces of molding. Working in a consecutive order means not having to ask yourself, Have I already measured that space? Use (+) or (−) symbols to represent ⅟₃₂ inch; this yields extremely accurate measurements, which produces tightly fitted joints.

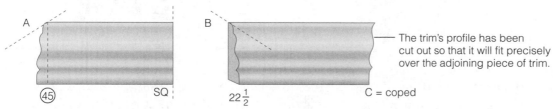

A B   The trim's profile has been cut out so that it will fit precisely over the adjoining piece of trim.

(45)   SQ   22½   C = coped

FIGURE 4-8 Example of how the recorded symbols show the angle of the cut on trim

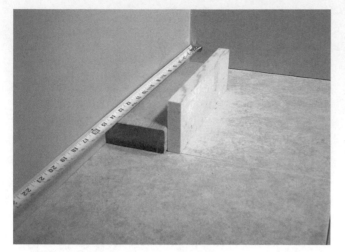

FIGURE 4-9   A block helps support the tape measure and keeps it from going beneath the sheetrock.

(a)

(b)

FIGURE 4-11   Scribing to fit

FIGURE 4-10   Taking an inside measurement, factoring in the length of the tape housing

FIGURE 4-12   Cutting an inch

Cutting an Inch

Some carpenters cut pieces of trim longer than needed so that they can set the pieces into place and scribe them to fit (**Figure 4-11**). *Cutting an inch* is a technique used when there is nothing to hook the end of the tape measure onto (as would be the case with an outside 45-degree cut). Line the tape up on the 1-inch mark and pull the desired measurement, adding 1 inch to it to compensate for the *cut inch*. For example, if you were measuring a piece that needed to be 12 inches long, your mark would be made at 13 inches if you were cutting an inch (**Figure 4-12**).

4-3 CUTTING BASE MOLDING

Once the measuring is completed, the focus shifts to the workstation where the trim will be cut. Miter saw stations are designed for the task of cutting all types of molding. Extendable rollers hold up the ends of

FIGURE 4-13 Miter saw table with extendable rollers

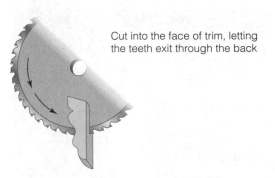

Cut into the face of trim, letting the teeth exit through the back

FIGURE 4-15 The saw blade should enter the face side of the trim and exit through the back.

longer molding, making it easier to measure and cut each piece (**Figure 4-13**). With flexible molding, the use of a support board underneath the molding will make cutting the trim easier.

In trim work, fractions of an inch make a tremendous difference in the quality of the finished product. A measurement of $\frac{1}{32}$ inch is only half of $\frac{1}{16}$ inch, but the difference between the two is still noticeable. To get an accurate cut on the miter saw, sight down the blade while it is turning (**Figure 4-14**). The blade can shift by fractions once the rpm has built up.

Cut through trim slowly. Fast cuts can damage trim, leaving jagged and splintered edges. The teeth of the saw blade should enter the face side of the trim and exit through the back side (**Figure 4-15**). **Figure 4-16** shows the damage caused by cutting trim from the back. Always hold trim securely against the base of the miter saw when cutting.

FIGURE 4-16 Damage caused from cutting trim from the back

4-4 COPING TRIM

Trim will need to be coped where two pieces meet at a 90-degree corner. The first piece, cut off squarely, fits

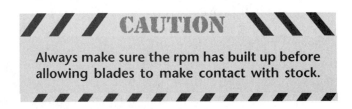

CAUTION

Always make sure the rpm has built up before allowing blades to make contact with stock.

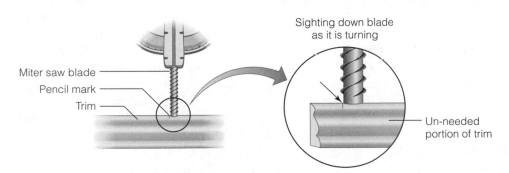

Sighting down blade as it is turning

Miter saw blade
Pencil mark
Trim

Un-needed portion of trim

FIGURE 4-14 Sight down the blade to get an accurate cut.

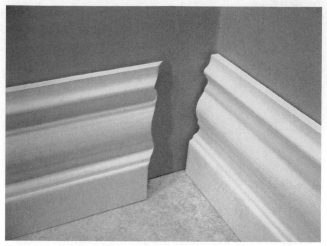

(a)

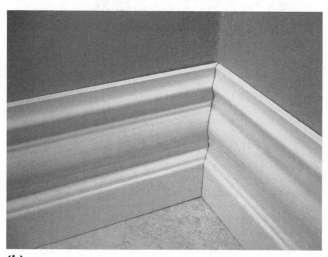

(b)

FIGURE 4-17   Coped trim fits over the previous piece.

FIGURE 4-18   An inside 45-degree cut ready for coping

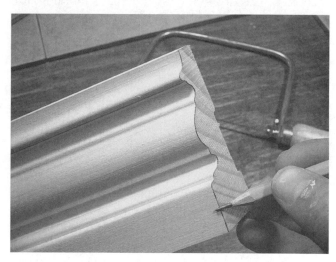

FIGURE 4-19   Trace the profile with the edge of a pencil.

all the way into the corner, whereas the adjacent piece is *coped* and will fit over the first piece (**Figure 4-17**). *Coped joints fit more tightly than mitered joints* and are more likely to remain tight over time, despite any settling or shifting of the structure. The coped piece more or less locks the end of the first piece into position. This is why it is important to cut and install the pieces in order. Mitered inside joints have a tendency to open up from pressure caused by nailing.

Coping is done using a coping saw, after the piece has received an inside 45-degree cut (**Figure 4-18**). Some carpenters prefer that the teeth of the blade angle back toward the handle of the coping saw, while others prefer to have the teeth facing the other direction. Either way is acceptable. The blade is thin and narrow so that the tight intricate profiles of trim can be cut with precision.

Before coping, it is helpful to trace the profile of the trim with the edge of a pencil (**Figure 4-19**). Doing this better defines the profile in a visual sense,

allowing you to cope without straining to see where the profile begins and ends. Prefinished trim, when cut for coping, has a definite profile that is easy to see.

Figure 4-20 shows a piece of trim as it is being coped. Take note of the blade angle in relation to the piece of molding. Molding such as base does not need a severe **undercut**, as crown does. The undercut is the amount of material removed from the back side of a piece as it is being coped. Holding the saw over 5 degrees is ideal when coping base and similar moldings.

Once the molding has been coped, the edges can be smoothed using sandpaper (**Figure 4-21**). Be sure to remove pencil marks from trim, especially stain-grade trim. Short pieces can be difficult to cope; it is helpful to cope the piece before cutting it to the final length. With flexible molding, instead of trying to cope the trim, leave the ends square and cope the pieces that are joining to it.

When coping trim, remember the following:

- Cope trim at inside 90-degree corners.

- Trace trim profile with a pencil after making an inside 45-degree cut.

- After coping, smooth the edges with fine-grit sandpaper and remove pencil marks.

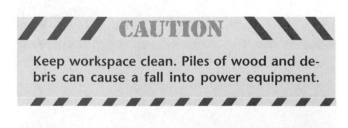

CAUTION

Keep workspace clean. Piles of wood and debris can cause a fall into power equipment.

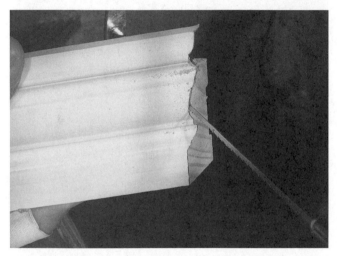

FIGURE 4-20   Trim as it is being coped.

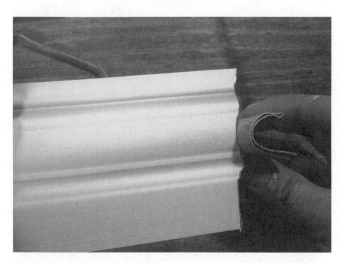

FIGURE 4-21   Fine-tune the edges of a cope with fine-grit sandpaper.

4-5 SPLICING TRIM

Splices should consist of 45-degree cuts and should occur over a stud so that a solid joint is created (**Figure 4-22**). Splices should be glued with wood glue. After splicing, any slight imperfections should be sanded smooth so that the finished product is seamless. An imperfect joint will be highly visible on the finished product.

Figures 4–23 through **4–25** show examples of uneven splices. Do not nail bad splices. Mark the piece

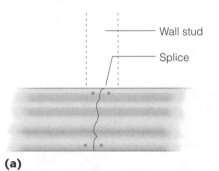

Wall stud

Splice

(a)

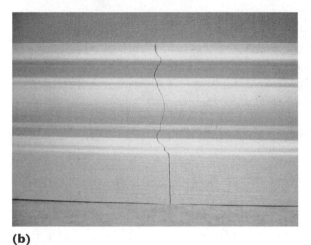

(b)

FIGURE 4-22   Splices should occur over a stud to create a solid joint.

FIGURE 4-23   Uneven splice

FIGURE 4-24   Uneven splice

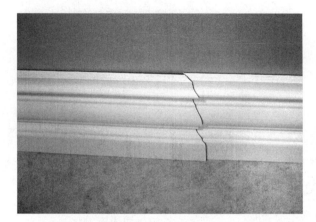

FIGURE 4-25   Uneven splice

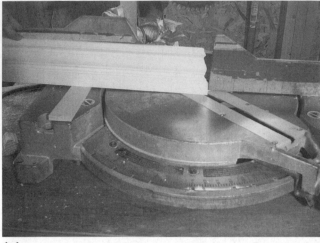

(a)

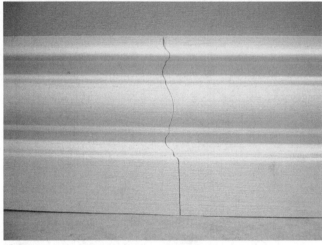

(b)

FIGURE 4-26   Shimming the base to correct a bad splice: (a) Place shim under molding to correct a bad splice; (b) A good splice.

and make the appropriate adjustment on the miter saw. Most splices can be adjusted by shimming up the base while recutting it on the miter saw (**Figure 4-26**). Doing this may shorten the piece by a fraction, but the next piece, a coped piece, will conceal the end of the slightly shortened piece (which is another reason why working in a specific direction is helpful).

> ### NOTE
>
> Trim pieces bought from two different sources will sometimes have been milled slightly differently, even though their profiles are similar, and they will not match up perfectly. If working with trim that is *slightly* mismatched, make changes in places where they will be less noticeable, such as at a doorway, rather than at a splice or in a corner. Trims that are greatly or obviously different should not be used in the same room. Using a "slightly different" trim in a small closet, however, is sometimes an option. ∎

4-6 INSTALLING BASE MOLDING

Start base installation with the piece designated as number one, installing the following pieces consecutively. If the finished floor is not installed, shim the base up anywhere from ⅜ to ⁷⁄₁₆ inch. (See the section later in this chapter on shims for base molding.) On finished floors, shims are not necessary; the base should rest firmly on top of the finished floor.

Locate studs before nailing. Studs are easily found at electrical outlets (**Figure 4-27**). Additional studs can be found by measuring every 16 inches. On walls without electrical outlets, locate a stud using a finish

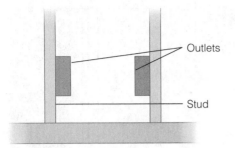

FIGURE 4-27 Studs can be located beside electrical outlets.

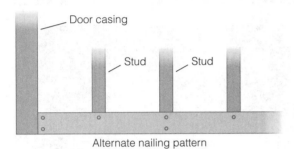

Alternate nailing pattern

FIGURE 4-28 Nailing pattern for base molding (top and bottom, alternating in between)

(a)

Top View of Casing

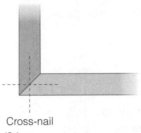

Cross-nail

(b)

(c)

FIGURE 4-29 Glue and cross-nail outside corners.

nail or a stud finder; and then measure to find the center of the others.

Check the fit of each piece of molding before nailing it in place. Use adequately sized finish nails for fastening trim. The recommended size nail for fastening base is a 16-gauge nail 2 inches in length. Nail at the top and bottom at beginning and ending points of the trim, alternating the nailing pattern in between as shown in **Figure 4-28**. Outside miters should be glued and cross-nailed (**Figure 4-29**). Use a small-gauge finish nail when cross-nailing to avoid splitting the wood. Try to stay at least ¾ inch away from the end of any trim when nailing it to avoid splitting it.

Sand outside corners after the glue has had time to set. On inside corners, use a small trim bar to pull the trim tightly into the cope and nail it while holding pressure on it (sometimes the impact from the finish nailer will cause the piece of trim to kick back, leaving a gap, if it is not supported by the trim bar). This will help to create a stable inside corner.

When fastening base near plumbing, use construction adhesive and 3/4- to 1-inch finish nails to avoid penetrating pipes. Use a nail set to countersink protruding nails (**Figure 4-30**).

FIGURE 4-30 Use a nail set to countersink nails.

When installing base molding, remember the following:

- Locate studs.

- Check the fit of each piece of trim before nailing.

- Shim base on unfinished floors.

- Install base on finished floors so it rests firmly on top of the floor.

- Use adequately sized finish nails.

- Cross-nail outside corners.

- Sand outside corners once glue has set.

- Around plumbing, fasten trim with construction adhesive, and use ¾- to 1-inch finish nails to avoid penetrating pipes.

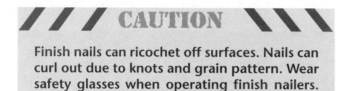

CAUTION

Finish nails can ricochet off surfaces. Nails can curl out due to knots and grain pattern. Wear safety glasses when operating finish nailers.

4-7 STACKING BASE MOLDING

The purpose of stacking trim is to achieve a larger and more attractive molding. **Stacked base** can make a room look rich and unique. The look that is achieved is different from what stores have to offer. Trim that cannot be produced by one milling can be made by the finish carpenter with multiple pieces of trim to give a greater dimensional effect. Another reason trim might be stacked is to seal the gap between the original base and tile or hardwood flooring (**Figure 4-31**). Stacked base, or *built-up*, moldings are stronger so that they can resist the effects of seasonal changes, more so than a single piece of molding could. Seasonal changes, changes in air temperature and humidity levels, can sometimes cause materials to slightly swell and contract, which occasionally results in joints and corners opening and caulk seams sometimes split open. This is especially apparent in homes that are not lived in year-round. Keeping heating and cooling systems on greatly reduces the effects of seasonal changes and changes in humidity levels.

FIGURE 4-31 The molding covers gaps between base and flooring.

Stacking trim is done simply by nailing one or more pieces onto one another. **Figure 4-32** shows some different profiles of stacked base molding. The techniques used to measure and record measurements of, cut, and install these added pieces of base are the same as the techniques already discussed for base molding, with a couple of exceptions. Generally these added pieces are smaller in dimension, and splices in smaller pieces can occur almost anywhere because there is something solid to which to nail the trim.

One thing to consider before installing extra pieces of base is figuring out how they will tie into door trim without appearing odd or out of place. Pieces of trim can be mitered (at 15, 22.5, or 45 degrees). Everything needs to flow and appear as though it were meant to be that way, and not be chopped off square, which creates an unattractive focal point in the room. **Figure 4-33** illustrates how stacked base can be tied into door trim.

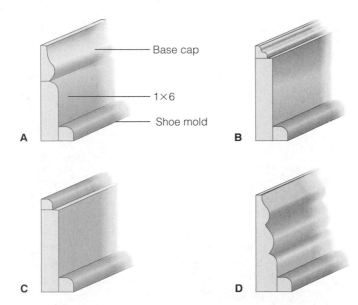

Base cap

1×6

Shoe mold

A   B

C   D

FIGURE 4-32 Profiles of stacked base molding

4-8 SHIMS FOR BASE MOLDING

Finish carpenters often begin their job before the flooring has been installed. Therefore, often the base molding has to be shimmed up so that later the flooring can be fit beneath the base (another option is to leave the base molding off until the floors are installed, but sometimes this creates a problem for the finish carpenter, especially if they have begun work on another job at another location). If you are installing base onto a finished floor, the molding should sit directly on the finished floor so that there are no gaps. On floors that are not finished (having no tile, hardwood, or carpeting), base molding is shimmed up so that the finished flooring can fit tightly beneath it. For carpeting and most tile, the base is shimmed up anywhere from ⅜ to ⁷/₁₆ inch (**Figure 4-34**). If you are using ¾-inch hardwood flooring, the base would be shimmed up ¾ inch or more so that the flooring can be fitted under or near the bottom of the base. Failing to shim the base results in a visual waste of trim. A 3-inch piece of base molding, if installed without shims, will appear as a 2½-inch piece after the flooring is installed.

Use the table saw or miter saw to cut shims of proper height. Cut several shims so that they can be alternated while installing the base molding (**Figure 4-35**). When installing the base molding, space shims 2 to 3 feet apart (or a comfortable arm span). Flooring contractors will later cut the bottom of the doorjambs so that they match the bottom of the base molding,

(a)

(b)

FIGURE 4-33 Tying base into door trim

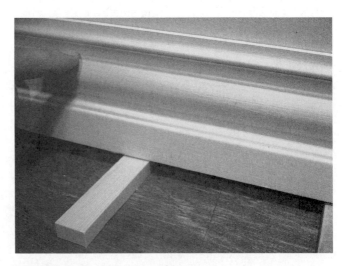

FIGURE 4-34 Shimming base

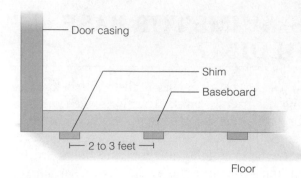

FIGURE 4-35 Space shims every 2 to 3 feet (or comfortable arm span).

FIGURE 4-36 Flooring fits uniformly under doorjambs and base molding.

and at that point the flooring will fit beneath everything uniformly (**Figure 4-36**).

After flooring materials have been installed, shoe molding is sometimes fastened to the bottom portion of the base to conceal small gaps and imperfections between the base and tile or hardwood flooring.

NOTE

Do not nail shoe molding to the finished flooring. Finished floors swell and contract due to seasonal changes and changes in humidity levels. This swelling and contracting will cause the shoe molding to pull away from the baseboards. Shoe molding should be nailed to the baseboards only. ■

MITER SAW POSITIONS AND SYMBOLS

Always practice safety when using the miter saw, keeping your fingers a safe distance away from the cutting path of the blade. When cutting small pieces of trim, use a backing board to prevent the piece from being damaged (**Figure 4-37**). A backing board will also help keep smaller pieces from being *launched* by the saw blade (the blade can launch pieces with enough velocity to possibly cause blindness if contact is made with an eye). Always wear safety glasses.

The miter saw comes with preset stops positioned at the most common cutting angles. Most saws are adjustable and can cut anywhere from 0 to 60 degrees, left and right. Compound miter saws are capable of cutting two angles at once (such as when cutting crown molding).

The photographs in **Figure 4-38** show the most commonly used angles when cutting trim and base molding. When you are recording measurements,

FIGURE 4-37 Use a backing board when cutting small pieces of trim.

you can use numbers and symbols to indicate which end of the trim receives what type of cut. The most commonly used miter positions are shown also in Figure 4-38.

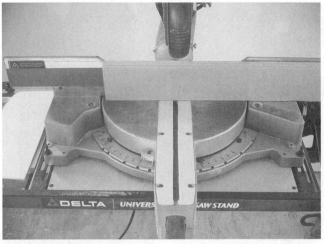

(a) Square cut

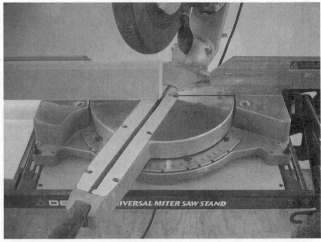

(b) Inside 22½-degree cut

(c) Inside 45-degree cut

(d) Inside 22½-degree cut

(e) Inside 45-degree cut

(f) Outside 22½-degree cut

FIGURE 4-38 Commonly used angles for cutting trim and base molding (miter saw positions).

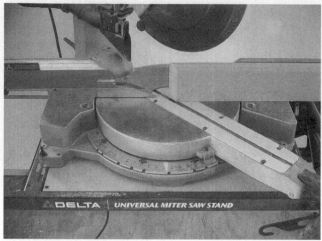

(g) Outside 45-degree cut

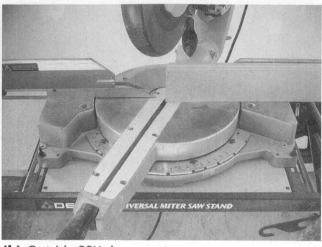

(h) Outside 22½-degree cut

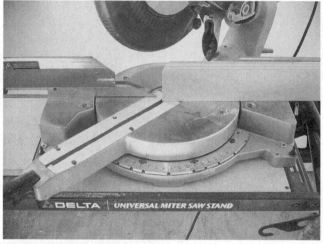

(i) Outside 45-degree cut

FIGURE 4-38 *(continued)*

SUMMARY

- Base molding is installed after doors have been installed and trimmed with casing.
- Most of the time, base is installed at a height that allows for flooring to fit beneath it.
- Floors should be free of debris and marked to indicate the location of plumbing before base is installed.
- Care should be taken when installing trim on walls where plumbing exists.
- Corners are not always a true 90 degrees.
- Cutting and installing trim in a specific order lessens confusion.

- Using numbers and symbols to record measurements makes dealing with a large number of pieces less confusing.
- Splices should consist of 45-degree angles and should occur over studs.
- All joints and outside corners should be glued.
- Uneven splices should be corrected before nailing.
- Shoe molding is sometimes used to close the gap between the base and tile or hardwood flooring.

KEY TERMS

cope To cut a trim's profile so that it fits precisely over an adjoining piece of trim.

miter To cut the end of a board at an angle; boards are usually mitered to meet with another board.

stacked base Multiple pieces of molding that are nailed together to appear as one molding.

undercut The amount of material removed from the back side of a piece as it is being coped.

REVIEW QUESTIONS

1. Which direction should one go in when taking measurements for base molding?

2. What is the best method for taking an inside measurement?

3. What symbols and abbreviations are used in recording base molding measurements?

4. Describe the technique of cutting an inch.

5. What tool is most helpful in locating the exact intersection of trim when trying retrieve a corner measurement?

6. Where should splices occur?

7. What angle should splices consist of?

8. Is it safe to say that corners are at true 90- and 45-degree angles?

9. How would you install trim behind a commode?

10. Is it safe to assume that plumbing pipes are protected by metal plates?

11. Which side of the trim should the saw blade enter?

12. What size shim should be used when installing base?

13. How would you determine the direct center of a stud?

14. What length of nail is ideal for nailing base?

15. Why is coping ideal for inside corners?

16. Before coping a piece of trim, what can be done so that the profile can be seen more clearly?

17. What can flexible trim be placed on while cutting it to make the process easier and safer?

18. When nailing trim, why is it a good idea to keep the nails at least ¾ inch away from the end?

19. How many feet of clearance should there be on either side of the miter table?

CHAPTER

5 Windows

OBJECTIVES

After studying this chapter, you should be able to

- Cut, assemble, and install window stools
- Make stools to fit any size window
- Cut aprons to fit beneath the window stools
- Build and install a complete window casing, with stool and apron

INTRODUCTION

Although many different sizes and styles of windows are available in today's market, windows used in interior finishing can be divided into two basic categories. The two main categories are *wood-cased windows* (**Figure 5-1a**) and *sheetrock return windows* (**Figure 5-1b**). ■

(a)

(b)

FIGURE 5-1   Window types: (a) Wood-cased window; (b) Sheetrock return window

TYPES OF WINDOWS

A **wood-cased window** (sometimes referred to as a *wood return window*) receives a **jamb** at the sides and top, is trimmed out with casing, and may have a **stool** at the bottom, beneath which is an **apron**. The stool is the bottom portion of window trim and it is connected to the jambs. The apron is the trim member installed beneath the stool. **Sheetrock return windows** have gypsum wallboard where jambs would otherwise be present. The two windows are similar in that they both receive window stools and get aprons beneath the stools. **Picture-framed windows** are similar to wood-cased windows, only they do not have a stool or apron at the bottom. On a picture-framed window, the jamb and the casing simply continue across the bottom, the bottom casing joining the sides with a 45-degree miter cut.

MATERIALS

Materials used for making jamb extensions should be of a solid wood product, vinyl, or PVC. Solid wood resists the small amount of moisture that is sometimes present around windows. Weather conditions and humidity can create moisture around windows, even high-quality windows that have been weather-sealed. Fibrous materials (particle board and other similar materials) should not be used (unless rated for exterior use), as they have a tendency to swell or warp when they come into contact with any type of moisture.

Window casing is fastened to the jambs and does not make contact with window frames. It is far enough away from the window so that there is little chance of it ever coming into contact with moisture. Therefore, window casings can be of either solid wood or fibrous material such as MDF.

PROCEDURES: FIT A WINDOW WITH JAMBS AND TRIM

 SAFETY TIPS

- Make sure ladders are secure before climbing up them.
- Use safety railings on scaffolds, and do not roll scaffolding over cords or hoses.
- Wear proper clothing. Baggy clothing and unbuttoned cuffs can get caught in power equipment.
- Make sure that the equipment switch is in the OFF position when plugging the cord into a power outlet, and make sure all tools and cords are grounded.

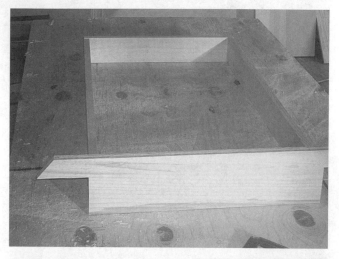

Following are the steps used in fitting a window with jambs and trim:

1. Measure the width needed for jamb stock.
2. Rip the jamb stock and remove saw marks. (Use fine-grit sandpaper to slightly round over sharp edges.)
3. Cut the jamb stock to the lengths needed.
4. Make the window stool.
5. Nail jambs and stool together, creating one unit.
6. Install the jamb unit using shims and finish nails.
7. Trim around the jambs with casing.
8. Cut and install an apron beneath the window stool. (A backing piece is helpful when cutting the small return pieces for aprons.)

FIGURE 5-2 Arrange jambs on worktable, and make sure pieces rest squarely against one another when nailing.

5-1 ASSEMBLING AND INSTALLING WINDOW JAMBS

Assemble jambs on an elevated surface, such as a worktable, to make the task less strenuous. Arrange the jambs and stool on the worktable, and fasten the pieces with finish nails. Make sure pieces are resting squarely against one another (**Figure 5-2**).

Before installing jambs, make sure to remove all debris (insulation, nails, joint compound, etc.) so that the jambs will sit tight against the window frame (**Figure 5-3**). Place shims between the rough framing and jambs to prevent any future movement (**Figure 5-4**). Keep a uniform reveal around the window frame. (Since the window itself is square, the frames can be set into place without using a level.)

Nail through the jambs at shim locations, slightly angling the nails to ensure a stronger hold. Use a nail set to set any nails that fail to countersink so that the nail heads are not protruding. Cut off protruding shims so that they do not interfere with the casing. Finish by adding the casing and the apron.

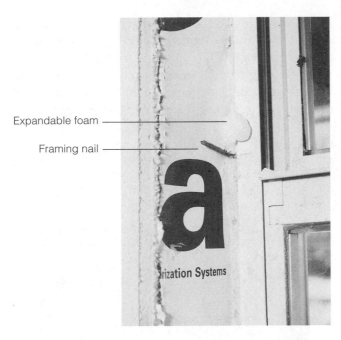

Expandable foam

Framing nail

rization Systems

FIGURE 5-3 Clear debris from rough opening before installing jambs.

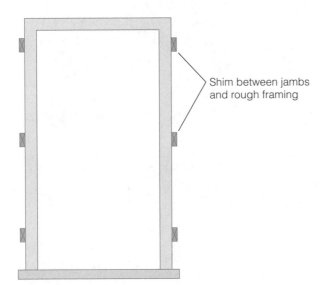

Shim between jambs and rough framing

FIGURE 5-4 Place shims around jambs to prevent future movement.

5-2 INSTALLING JAMB EXTENSIONS

Jamb extensions extend from the window frame and end up being flush with the wall surface. Having jambs that are flush with the wall surface makes it possible to add trim, which will conceal the space between the jambs and the wall framing (**Figure 5-5**). Windows can receive jambs on all four sides; however, most carpenters prefer to have a defining element, such as a window stool, at the bottom of the window.

FIGURE 5-5 Trim conceals the gap between the jamb and rough framing.

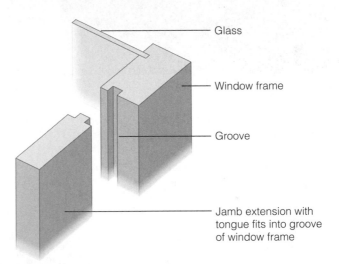

Glass

Window frame

Groove

Jamb extension with tongue fits into groove of window frame

FIGURE 5-6 Some windows have grooved channels made for accepting a tongue on factory-made jambs.

Some custom-ordered windows come with their own set of extension jambs that have already been cut to accommodate the specified wall thickness. In this case, the windows will have receiver channels milled into the frame. A **receiver channel** is a groove designed to accept the tongue of the jamb extension (**Figure 5-6**). Jambs for custom-ordered windows come precut and ready to assemble.

Making Custom Jamb Extensions

To custom-make extension jambs, the width needed for the jambs must first be determined by measuring from the face of the window to the wall surface (**Figure 5-7**). Take measurements at several different locations around the window to see if the measurements are all the same. (Sometimes the measurements will fluctuate a little.) **Rip**, or cut, material for jambs on the table saw.

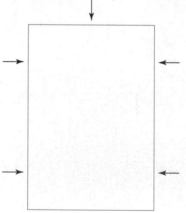

FIGURE 5-7 Measure for the width of the jamb stock at several points around the window.

Rip the jamb stock ⅟₁₆ inch larger than the size needed so that the material can be trued on a jointer or planed down to remove the saw marks. Later, if one of the jambs extends out too far, the jamb can be block-planed until it is flush with the wall surface.

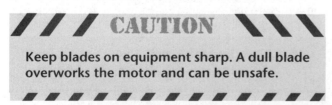

CAUTION

Keep blades on equipment sharp. A dull blade overworks the motor and can be unsafe.

Once a finished width has been achieved, the jambs can be cut to the length needed. Length is determined by measuring the window frame. When measuring the window frame, remember that the

jambs will have to be cut long enough to leave a **reveal** around the window frame. A reveal is the amount of offset between two components (as discussed in Chapter 3). Once the desired amount of reveal has been determined (⅛ to ¼ inch), make allowances for it as illustrated in **Figure 5-8**. Looking at the figure, take

(a)

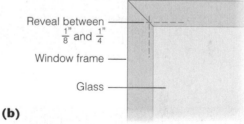

Reveal between $\frac{1}{8}"$ and $\frac{1}{4}"$

Window frame

Glass

(b)

(c)

Allow for reveal

FIGURE 5-8 Allowing for the reveal (amount of offset between face of window and jamb):
(a) Marking reveal; (b) Reveal should be ⅛ to ¼ inch;
(c) Reveal has been allowed for in this window.

notice of the following:

- The head jamb has been cut with allowances made for the reveal and for extending over the side jambs (the length of the head jamb will be used in calculating the size of the stool).

- The side jambs have been cut with allowances made for the reveal.

- The jambs are flush with the wall surface.

- The jambs sit tight against the window frame.

Once the jambs have been cut to the proper lengths, a stool can be made so that all the components (side jambs, head jamb, and stool) can be nailed together to form a unit. This makes the installation easier and creates a finished product that is more professional-looking than one that is installed one piece at a time (a method that involves too many variables and allows for too many mistakes to occur).

5-3 INSTALLING WINDOW STOOLS

The width of the window stool has to be wider than the jamb stock so that the window casing will have something to rest on (**Figure 5-9**). In most cases, the width of the window stool is the width of the jamb stock plus twice the thickness of the casing being used (**Figure 5-10**). The total length of the

FIGURE 5-9 Window casing rests on extended stool.

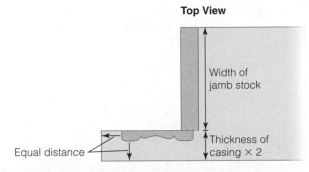

Top View

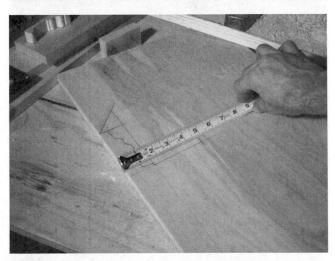

FIGURE 5-10 Determining the width of the window stool

window stool is calculated by adding the length of the head jamb and the casing widths, plus the reveal and twice the casing thickness. To simplify the procedure, the window stool can be laid out as shown in **Figure 5-11**. Notice the reveal and the casing, and notice that the overhang is the same as the thickness of the casing.

After marking the stool, you can cut out the unneeded portion using a jigsaw or a table saw in

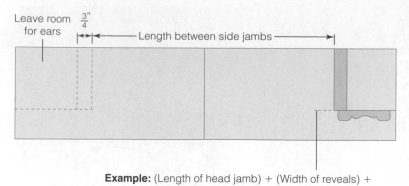

Leave room for ears $\frac{3''}{4}$

Length between side jambs

FIGURE 5-11 Laying out the window stool to find the length

Example: (Length of head jamb) + (Width of reveals) +
(Width of casing × 2) + (Thickness of casing × 2)

FIGURE 5-12 Cutting out the unneeded portion of the window stool

combination with a miter saw (**Figure 5-12**). After you've made the cuts, the stool has what is referred to as either **ears** or **horns**. The ears, or horns, extend out on either side. Casing rests on top of the ears.

Often the edges of the window stools are shaped using a router (**Figure 5-13**). Routing the edges of the stools can be done before the unneeded portion is cut out to avoid damaging the wood (sometimes a router bit splinters the wood as it exits). Routing the piece in a counterclockwise direction will also reduce damage to the wood (**Figure 5-14**). Use a sharp router bit, and do not overwork the router. Overworking the router

FIGURE 5-13   Shaping the edges of a stool using a router

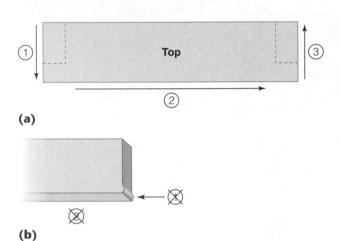

(a)

(b)

FIGURE 5-14   Routing in a counterclockwise direction: (a) Preferred method of routing; (b) Routing sides out of order could damage finished edge.

can cause **chatter**, which occurs when the bit vibrates against the wood, leaving what appears as grooves or ridges.

Sometimes trim is added to the edges of the window stools (**Figure 5-15**). **Figure 5-16** shows some different profiles for window stools. Ears are returned back to the wall at either a 45- or 90-degree angle, and sometimes the ears are rounded over (**Figure 5-17**).

Window stools for sheetrock return windows are made basically the same way with the exception of the ears. A stool for a sheetrock return window does not have to allow for casing or side jambs; therefore, the ears are shortened until a balanced look is achieved (**Figure 5-18**).

After the stools and the jambs have been cut as needed, they can be fastened together and installed as one unit.

(a)

(b)

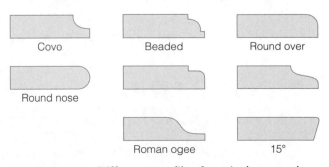

(c)

FIGURE 5-15   Adding trim to the edge of a window stool

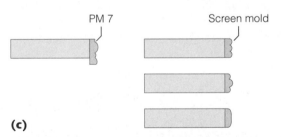

FIGURE 5-16   Different profiles for window stools

Top View

Window stool

90° return   45° return   Round return

FIGURE 5-17   Returning the ears of the window stool

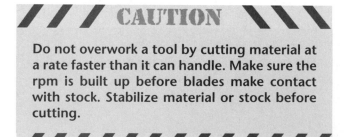

(a)

Top View

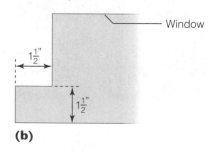

Window

$1\frac{1}{2}"$

$1\frac{1}{2}"$

(b)

FIGURE 5-18   Balancing the look of a sheetrock return window stool

(a)

Head casing

Scribe side casing length

Rest point of casing on stool

Stool

(b)

FIGURE 5-19   Steps in cutting and installing window casings

5-4 INSTALLING WINDOW CASINGS

There are many different types of window casings. Casing used on the windows normally matches the casing used on the doors so that continuity is maintained. Cut the number of casings needed for all the windows with left- and right-hand miters. Cut casings long, and scribe the length of the side casings after the head casing has been installed (**Figure 5-19**).

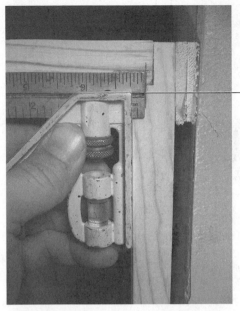

Reveal marks (where edge of casing will sit)

(a)

(b)

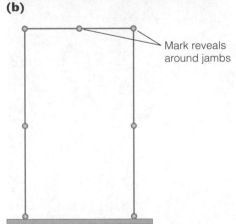

Mark reveals around jambs

(c)

FIGURE 5-20   Marking reveals onto jambs with a combination square

Use glue on all miter joints, and make sure the joints fit tight. Keep a uniform reveal around the jambs when installing casings. Mark reveals on jambs using a gauge block, a combination square, or a tape measure (**Figure 5-20**).

Rosettes, proportional to the size of the casing, are sometimes used at the upper corners of windows, where the head casing and side casing meet. The casing joins the rosette with a square cut (**Figure 5-21**). Windows trimmed with square trim are sometimes joined at the corners with a square cut instead of a miter cut (**Figure 5-22**), with the head casing extending over the side casings.

FIGURE 5-21   Casing joins rosette with a square cut

Square trim sometimes joins with square cuts

FIGURE 5-22   Windows with square trim are sometimes joined with a square cut

5-5 INSTALLING WINDOW APRONS

Aprons fit directly beneath the window stool. Typically, the apron is the same material used for the window casing. Each end of the apron is in line with the outside edges of the window casing (**Figure 5-23**). On sheetrock return windows, the edges of the apron are in line with the edge of the wallboard (**Figure 5-24**).

The ends of the apron are often returned back to the wall by cutting small return pieces, as shown in **Figure 5-25**. If cutting smaller return pieces, use a backing board for safety (as shown in the figure). Wood glue is often enough to secure the return pieces to the apron. Some carpenters bypass using the return pieces by cutting a slight angle, 15 degrees in most cases, at the ends of the apron (**Figure 5-26**).

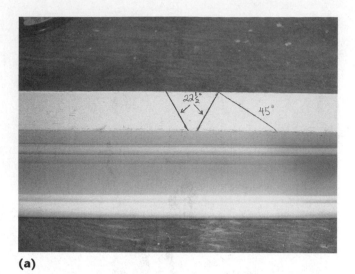

(a)

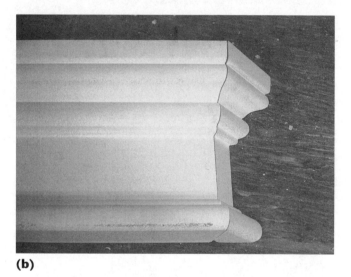

(b)

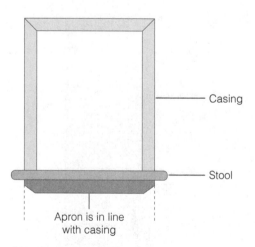

FIGURE 5-23 Ends of the apron are in line with the edges of the casing

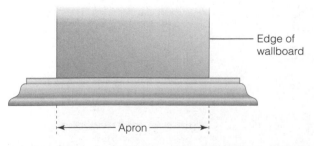

FIGURE 5-24 Ends of the apron are in line with the edge of the wallboard on sheetrock return windows

(c)

FIGURE 5-25 Returning window apron to the wall: (a–c) 45-degree returns (d–e) 90-degree returns

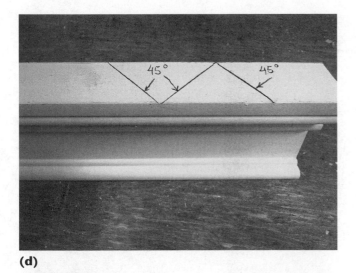

(d)

(a)

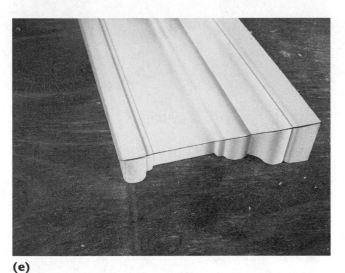

(e)

FIGURE 5-25   (continued)

Stool

Apron

15° cut intersects with 90° cut

(b)

FIGURE 5-26   Cutting a 15-degree angle onto apron

Recording Measurements for Window Stools (Sheetrock Return Windows)

It is helpful to use a diagram when recording measurements for window stools (**Figure 5-27**). Learning a method for measuring and recording window stool measurements results in efficiency. When recording measurements for stools, pick a window to begin with and designate it as number one. Assign numbers to the diagrams so that they correspond with

Window #1

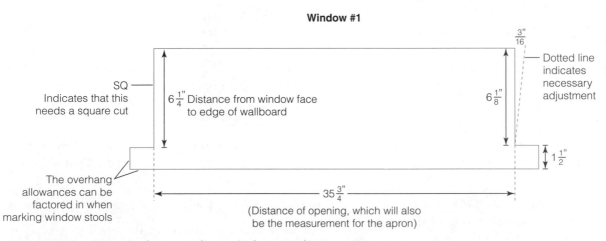

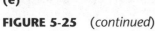

$\frac{3"}{16}$

Dotted line indicates necessary adjustment

SQ
Indicates that this needs a square cut

$6\frac{1}{4}"$ Distance from window face to edge of wallboard

$6\frac{1}{8}"$

$1\frac{1}{2}"$

The overhang allowances can be factored in when marking window stools

$35\frac{3}{4}"$

(Distance of opening, which will also be the measurement for the apron)

FIGURE 5-27   Diagram for recording window stool measurements

the correct window. Go from window to window, moving about the house in a clockwise or counter-clockwise direction, and record all window stool measurements.

The following are steps to recording measurements for window stools:

1. Hold a speed square against the face of the window and determine whether or not the sheetrock is square in relation to the window face. If the sheetrock isn't square in relation to the face, or frame, of the window, mark the needed amount of adjustment on the diagram.

2. Measure from the face of the window to the edge of the wallboard and record the measurement on the diagram.

3. Take an inside measurement at the front of the window opening. A measurement does not need to be taken at the back using this method.

4. When marking window stool material with corresponding measurements, add overhang allowances as shown in the illustration.

5. After cutting out the unneeded portion from the stool material with a jigsaw, number the stools so that they match up to the proper window.

SUMMARY

- Sheetrock return windows receive a stool and apron only.
- Real wood products are preferred when it comes to window jambs and stools, as they resist moisture better than fibrous manmade materials.
- Use fine-grit sandpaper to slightly round over sharp edges of jambs.
- A backing piece is helpful when cutting the small return pieces for aprons.
- Rough openings around windows should be cleared of any obstructions before you attempt to install any stools or window casements.

- Windows are not always centered perfectly in rough openings.
- Common reveals for windows range from ⅛ to ¼ inch.
- Window stools can be routed with many different profiles.
- Routing stools in the proper direction decreases the chances of damaging edges.
- Reveals on window casings should be kept uniform all the way around.

KEY TERMS

apron Trim member installed beneath the window stool; usually made with the same trim that is used for the window casing.

chatter Ripples or grooves left behind in the wood when a router bit vibrates against the material, caused by using a dull router bit or overworking the router.

ears Sometimes referred to as horns; the part of the stool that extends out on either side of the window opening; casing rests on top of the ears.

jamb Vertical piece forming the side of an opening; usually ¾-inch thick; extends from the face of the window and ends up being flush with the wallboard; casing is nailed to the jambs.

picture-framed windows Window that have a jamb and casing on all four sides; do not have a stools or aprons.

reveal Amount of offset between two components.

receiver channel A groove designed to accept the tongue of a jamb extension.

rip Ripping is done using a table saw; material is *ripped* to the width needed.

sheetrock return windows Windows that are cased with wallboard; most sheetrock return windows receive a stool to give the bottom of the window a defining element.

stool Bottom portion of window trim; connected to the side jambs.

wood-cased window A window that is cased with wood on all four sides (jambs, casing, stool, apron).

REVIEW QUESTIONS

1. What things should be considered as possible obstructions in rough openings around windows?

2. When installing window frames, what can you use instead of a level to judge their correctness? In other words, what would you look at around the window to make sure it is the same all the way around?

3. Several methods can be used to mark reveals around window frames. What are two of these methods?

4. The ears of window stools can be returned back to the wall with what methods?

5. What are some common reveals around window frames?

6. If the reveal at the bottom of a window frame is ³⁄₁₆ inch, what should the reveal at the top be?

7. Why is it important to rout something in a particular direction?

8. What must be done before stools or frames can be slid into place?

9. What component fits directly beneath the window stool?

10. Return pieces for aprons can consist of 45- or 22½-degree pieces. Since a nail could possibly split the small pieces, how can they be fastened?

11. What purpose does shimming window jambs serve?

12. What can be done when cutting the small return pieces for aprons to ensure that the task will be done safely?

13. Like doors, windows can receive a decorative piece at the top corners. What are these corner pieces called?

14. If the reveal on the window jamb is ¼ inch along the top, what should the reveal along the side jamb be?

15. What type of material is commonly used for making window jambs and stools?

Wainscot, Picture Molding, and Chair Rails

OBJECTIVES

After studying this chapter, you should be able to

- Establish layout lines in preparation for wainscot
- Cut, assemble, and install materials involved in wainscot
- Lay out and install picture molding and chair rails
- Solve problems in installing wainscot around windows and doors

INTRODUCTION

Wainscot is a wall treatment that covers the bottom portion of a wall, usually at a height of 32, 36, or 60 inches. The 32-inch wainscot is typically used on 8-foot walls. The 60-inch wainscot is a style referred to as **mission wainscot**.

Wainscot can be assembled as one unit or piece by piece. It is applied after a level line has been marked onto the wall using a level and chalk line or a laser level. A cap rail is usually installed along the top, and a baseboard is placed along the bottom; this style is called *bead board* wainscot. ■

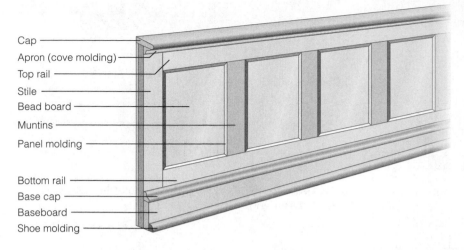

Cap
Apron (cove molding)
Top rail
Stile
Bead board
Muntins
Panel molding

Bottom rail
Base cap
Baseboard
Shoe molding

FIGURE 6-1 Frame and panel wainscot

TYPES OF WAINSCOT, PICTURE MOLDING, AND CHAIR RAILS

Frame and panel type wainscot is made up of **stiles**, **rails** and **muntins** joined to form rectangles on top of paneling (**Figure 6-1**). A stile is an outside trim member, such as on wainscot, cabinet face frames, doors, and so on. Stiles usually run vertically from the top to the bottom; their height is not shortened by a rail. Rails are horizontal members, such as on wainscot, cabinet face frames, doors, and so on. Rails begin and end at a stile. Muntins are vertical trim members in between the rails. Muntins, in most cases, are shorter in length than stiles because they run in between the top and bottom rails. **Cove molding** or a similar molding is usually installed along the inside edges of the rectangular frames on wainscot. Cove molding is a smaller trim with a concave profile.

The frame type of wainscoting involves more preliminary planning on the carpenter's part in order to calculate rectangles of equal size and spacing. The ideal size for a rectangle calls for one side to be 1½ times larger than the adjacent side (**Figure 6-2**). Achieving this ideal size may not always be possible for various reasons; however, one should attempt to get close to it.

Tongue-and-groove material is used for some styles of wainscot; it is applied with the same techniques used in bead board applications. There are several different styles of wainscot:

- Bead board wainscot (**Figure 6-3**)

- Frame and panel wainscot (as shown in Figure 6-1)

- Tongue-and-groove wainscot (**Figure 6-4**)

Ideal rectangle

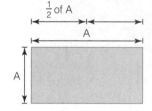

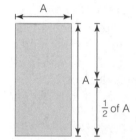

FIGURE 6-2 The ideal size for a rectangle

FIGURE 6-3 Bead board wainscot

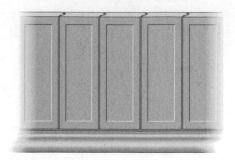

FIGURE 6-4 Tongue-and-groove wainscot

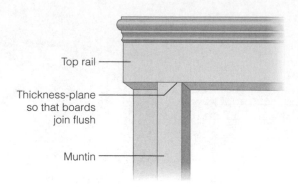

Top rail

Thickness-plane
so that boards
join flush

Muntin

FIGURE 6-5 Make sure to thickness-plane boards so that they join flush.

MATERIALS

Materials used for frame and panel wainscoting should ideally be of a solid wood (pine, poplar, oak, etc.). Solid wood can be joined with various techniques similar to the joinery techniques used in cabinetry. (See the section later in this chapter on assembling wainscot.) Some manufactured materials and fiberboards cannot be joined with the methods necessary for creating strong joints that will last over time.

Stiles, rails, and muntins should consist of a solid wood for joining purposes. Other components of the wainscot (such as the panel inserts, base, and cap rail) can be constructed of fiberboard if a paint finish is desired. If a stain-grade finish is desired, *all* components should be of the same species of solid wood (pine, oak, birch, ash, mahogany, etc.).

For all pieces that are going to be joined (stiles, rails, and muntins), make sure the pieces have been thickness-planed so that the joining boards will be flush with one another (**Figure 6-5**). Use a palm or orbital sander to sand any imperfect joints flush.

CAUTION

Use tools and machines for their intended purposes only, and pay strict attention when operating them. Make sure safety features are working properly, and keep tools and machines clean to avoid a buildup of sawdust and other materials that could affect the safety features.

PROCEDURES: CONSTRUCT WALL FRAMES AND WAINSCOTING

Revisit the following safety tips (introduced in Chapter 2) when operating the saw:

 SAFETY TIPS

• Keep blades on equipment sharp. A dull blade overworks the motor and can be unsafe.
• Unplug tools before changing blades or performing any other kind of work on them.
• Do not remove or pin back guards on tools.

• Inspect trim and make sure there are no hidden nails or staples present before making cuts.
• Stabilize material or stock before cutting, and make sure saw base sits flat on the material being cut.
• Adjust blade depths properly (usually ⅛ to ¼ inch more than the thickness of the stock that is being cut).
• Make sure the rpm is built up before blades make contact with stock.

- Cut material so that the cutting action of the tool is away from the body.
- Use a backing board when cutting smaller pieces of trim on the miter saw. A backing board will help prevent smaller pieces from being launched from the saw (which can cause serious injury to anyone standing nearby).
- Use push sticks to keep hands away from cutter heads and blades.
- Do not overwork a machine by cutting material at a faster rate than the machine can handle.

6-1 CONSTRUCTING AND SPACING WALL FRAMES

Wall frames are constructed using picture molding or a similar type of molding (**Figure 6-6**). The frames should be plumb and square, and the spacing between them should be made even. Use a laser level or chalk lines to establish a level line for the bottom course of picture mold. Use a level to make a plumb line for each square or rectangle (**Figure 6-7**).

The molding, if cuts are accurate, should join together and form a perfectly square unit. Vertical sections of molding may fall between studs and therefore will need to be attached to the wallboard with construction adhesive and finish nails. A chair rail is sometimes incorporated with wall frames (**Figure 6-8**). The chair rail should also be installed along a level line. Chair rail is usually installed at a height of 36 inches.

FIGURE 6-7 Use a level to make plumb lines for picture mold frames.

FIGURE 6-8 Chair rail

FIGURE 6-6 Wall frames are constructed with picture molding or material of a similar type.

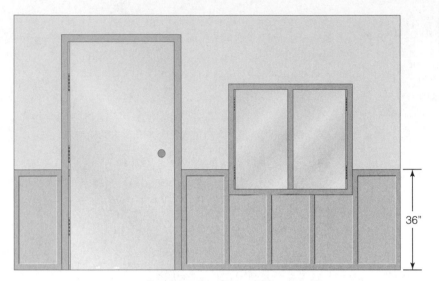

FIGURE 6-9 Wall drawn on graph paper

Spacing Frames

Spacing should be calculated for each wall section. Frames should be consistent in size along each wall section. To calculate the spacing, prior to making the panels, it is helpful to make a scaled-down drawing of the room on graph paper. Doing so can save you valuable time in the long run because it allows you to get an idea of how things will work out before beginning any actual work. Potential problems, such as working around windows, can be planned for in advance, so there will be no surprises. (See the section later in this chapter on meeting window and door trim.)

Figure 6-9 shows a wall that has been drawn on graph paper. Once the measurements have been taken, a drawing can be done on graph paper so that a plan is developed to ensure that everything will go smoothly.

Height of Frames

It is easier to determine vertical spacing than to calculate frame width, since vertical spacing dictates the spaces between frames. To properly lay out the vertical spacing of the wainscot, draw a diagram, starting at the top and working down. Include the dimension of each piece in the drawing. For example, suppose the wainscot is to be at a finished height of 36 inches. **Figure 6-10** shows the dimension of every piece that would be needed. This diagram gives you some idea about the options you have as far as height is concerned.

Typically, whatever size rail you decide upon at the top, you will need to have the same size for the

bottom, as well as between the frames, the stiles, and the muntins. So, for example, if you are using 3 inches of exposure space for the top and bottom rail, the muntins in between the frames will need to be 3 inches also. Keep the stiles and muntins a consistent width on all walls. The top rail can be cut at 3¾ inches (¾ inch of it will be covered by the cove molding) so that there still will be 3 inches of exposure after adding cove molding beneath the cap rail. The bottom rail needs to be exposed 3 inches also, so the height of the baseboard and base cap must be factored in (**Figure 6-11**). Remember that on an *inside corner* (opposite of an outside corner), one of the stiles will need to be wider

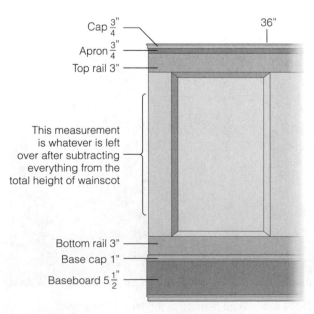

FIGURE 6-10 Dimensions of wainscot

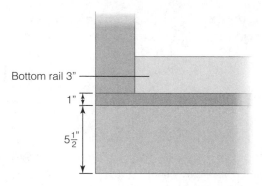

FIGURE 6-11 Factor in the height of the baseboard and base cap.

Top View

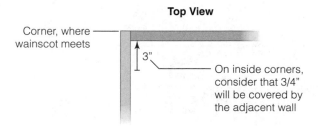

Corner, where wainscot meets

3"

On inside corners, consider that 3/4" will be covered by the adjacent wall

FIGURE 6-12 A stile on the inside corner

than the others, since ¾ inch of it is going to be covered by the adjoining stile (**Figure 6-12**).

Horizontal Spacing for Frames

To calculate the size for the individual frames, first measure the length of the wall. Then add up the number of spaces, or stiles or muntins, as shown earlier in Figure 6-9. To get a count of these, you will have to take an educated guess as to the number of frames you are going to end up with. This is where graph paper is useful. By looking at the vertical breakdown of how the wainscot is laid out, we can see that after subtracting the base, base cap, top cap, trim under the cap, and the top and bottom spacing, we are left with 22½ inches. If we are using a vertical frame, we already know that the width of that frame will be somewhere in the vicinity of 15 inches (because the ideal size for a rectangle is for one side to be 1½ times larger than the adjacent side). If we were going to use a horizontal frame, we would know that our width would be somewhere in the vicinity of 33¾ inches, simply because 22½ times 1.5 is 33¾ inches. We still do not know the exact size at which our frames will be, but having this general idea will assist us in calculating the exact size because now we can

guess at how many individual frames there will be, which allows us to count the number of spaces (stiles and muntins).

Now take the length of the wall and subtract the combined widths of the stiles and muntins. (Remember that these should be the same as the top and bottom rails, which are 3 inches.) Then divide that number by the number of spaces. Doing that calculation gives the exact distance needed between the stiles and muntins. These same calculations can be used for figuring the spacing of wall frames or picture mold frames. For example, a 20-foot wall is 240 inches. If we knew that we wanted spacing between muntins to be around 15 inches, and that the muntin was going to be 3 inches, we could divide the space of the wall by the 15-inch space plus the 3-inch width of the muntin, which would give us 18 inches. The calculation would look like this:

$$240" \div 18" = 13.333333"$$

Since a whole number is needed, we would round to 13. That means there would be room for 13 frames, which means we would end up with a total of 14 stiles and muntins.

14" × 3" (width of stiles and muntins) = 42"

240" − 42" leaves us with 198" of actual space

198" ÷ 13 (# of frames) = 15.230769, which is rounded to 15¼"

6-2 ASSEMBLING WAINSCOT

As mentioned earlier, wainscot can be assembled as one unit, or it can be installed piece by piece. On some walls, especially the ones that contain windows, you may want to first install the bead board if that is what you are using. Use paneling adhesive and place the top of the bead board or paneling ¾ inch below the desired finished height. (The ¾-inch cap will bring it up to the correct height.)

Whether using a standard level or a laser level in preparation for the wainscot, you may want to make the reference line ¾ inch lower than the desired height. Doing this will keep you from having to make allowances for the cap. If using bead board, make sure to begin the first panel where the first groove matches the last groove. This way the grooves should appear uniform once the frames are put on.

At some point, stile and muntin materials will need to be ripped on the table saw at the desired width. Make sure to allow enough room so that the material can be planed or sanded to get rid of saw marks. Make sure the thickness of all the pieces is the same; otherwise, there will be a noticeable difference at joint locations.

Once the bead board is on, work can begin on the frames. Calculating the spacing of the muntins is the first thing that needs to be done. To do this, measure each section of wall that is being wainscoted. Once the spacing has been determined, cut the top and bottom rails to the proper length. Lay out both rails at the same time, marking the center of each muntin (**Figure 6-13**).

Some finish carpenters prefer to fasten the frames together using a biscuit joiner (**Figure 6-14**). A slot is cut into the rail and the muntin, where a biscuit is glued, holding the two together. A clamp is placed at each joint (**Figure 6-15**). Follow the recommended drying time to allow the glue to set properly.

Other types of joinery systems can be used to join the components of wainscot, some of which are used by cabinetmakers. A doweling jig and dowels can be used to join frames. Pocket screw joinery, a face frame tool, utilizes a jig to cut holes for screws, which secures the frame at acute angles (**Figure 6-16**).

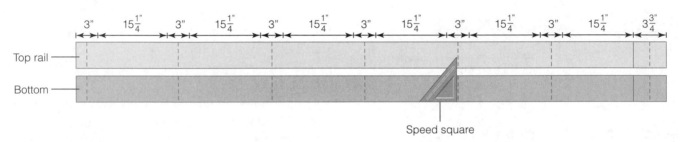

FIGURE 6-13   Laying out both rails at the same time

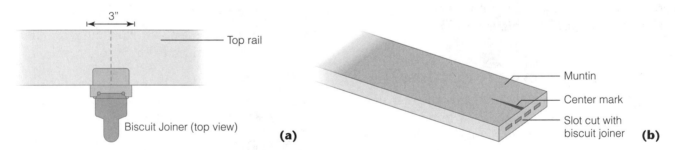

FIGURE 6-14   Fastening frames with a biscuit joiner

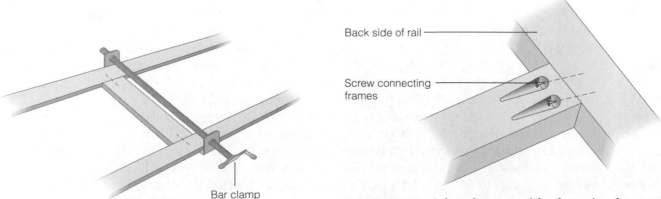

FIGURE 6-15   Clamp frames at each joint

FIGURE 6-16   A face frame tool for fastening frames together

Once the frames have set long enough, each one can be nailed into place, even with the top of the bead board, which is ¾ inch below the desired finished height (**Figure 6-17**). A cap can be nailed onto the top rail (**Figure 6-18**). A molding, such as cove, can be nailed beneath the cap, enhancing appearance while providing stability. Cut and install the base molding and a base cap. Install cove molding around the inside perimeter of the frames (**Figure 6-19**).

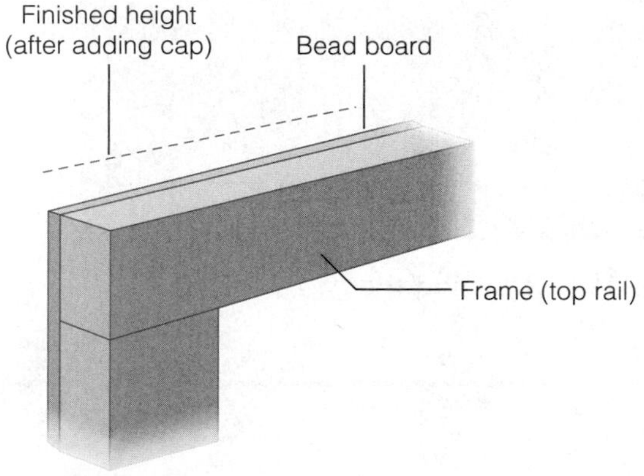

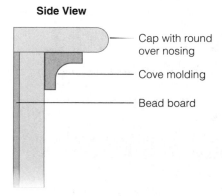

FIGURE 6-17   Nail frames even with the top of the bead board

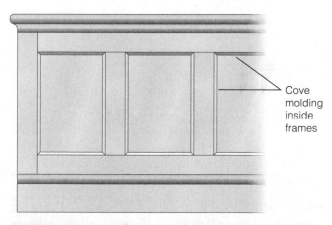

FIGURE 6-18   A cap on top of the top rail

FIGURE 6-19   Install cove molding to inside of frames

6-3 MEETING WINDOW AND DOOR TRIM

One of the problems to consider when installing wainscot is how to handle the areas where the wainscoting will butt into windows and doors. The thickness of the bead board combined with the ¾-inch thickness of the wainscot material makes the wainscoting thicker than most window and door trim. To remedy this, simply rout a transitional edge where the wainscot meets the window or door trim. Use a beaded edge, a cove, or whatever fits the situation (**Figure 6-20**).

This problem does not occur where picture mold and wall frames are concerned. On wainscot caps and chair rail, where meeting with window and door trim, the piece can be ended by cutting it at a 45-degree angle, as shown in **Figure 6-21**. Or the piece can be cut in such a way that it returns into the door or window trim, where either a 45- or 22½-degree angle is cut, and then a small return piece is cut and glued into place (**Figure 6-22**).

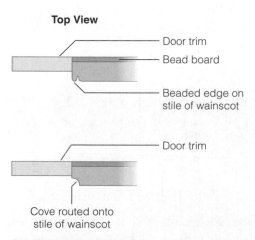

FIGURE 6-20   Transitional edges meeting windows and doors

Top View

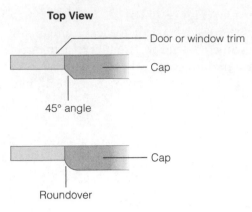

Door or window trim

Cap

45° angle

Cap

Roundover

FIGURE 6-21   A cap ending at window or door trim

FIGURE 6-22   Returning the cap into door or window trim

SUMMARY

- Wainscot is usually at a height of 32, 36, or 60 inches.
- The 32-inch wainscot is typically used on 8-foot walls.
- The height of the wainscot referred to as mission style is 60 inches.
- Wainscot can be assembled as one unit or piece by piece.
- A level reference line is needed when installing wainscot.
- In the ideal rectangle, the golden rectangle, one side is 1.5 times larger than the adjacent side.

- Spacing of wainscot and picture frames must be planned and calculated correctly.
- Graph paper is a useful tool when planning for wainscot. Story poles are also useful for planning.
- It is easier to determine vertical spacing than to calculate frame width, since vertical spacing dictates the spaces between frames.
- To solve problem areas, such as around doors and windows, a transitional edge can be routed onto the edge of the wainscot.

KEY TERMS

cove molding Smaller trim with a concave profile.

mission wainscot Wainscot that has a height of 60 inches.

muntins Vertical trim members that exist between top and bottom rails.

rails Horizontal trim members found on wainscot, cabinet face frames, doors, and so on.

stiles Vertical trim members; a stile lies at the outer edges of doors, cabinet face frames, or wainscot, so its height is not shortened by rails.

REVIEW QUESTIONS

1. Why is a level reference line important when installing wainscot?

2. What particular style of wainscot has a height of 60 inches?

3. What can be done where wainscot meets doors or windows?

4. Is there a predetermined size for stiles and muntins?

5. What is an ideal rectangle?

6. What would be more proportional for a room with 8-foot ceilings: vertical frames or horizontal frames?

7. How is spacing calculated?

8. Can frames of equal size and spacing be achieved without prior planning?

9. What is one method of joining the top and bottom rails with the stiles and muntins?

10. Is 32 inches an acceptable height for wainscoting?

11. Why should materials used for frames be thickness-planed?

12. Should paneling adhesive be used when installing bead board?

13. What is a suitable material for constructing frame and panel wainscoting?

CHAPTER 7 Closets

OBJECTIVES

After studying this chapter, you should be able to

- Mark layout lines for shelves and shoeboxes
- Make and install shoeboxes
- Cut and install support systems
- Make and install closet rod supports and rods
- Cut and install shelves

INTRODUCTION

This chapter discusses the basic closet components shown in **Figure 7-1**. With knowledge and understanding of how to build a basic closet, you will easily be able to perform the task of completing any closet design, regardless of its complexity. ■

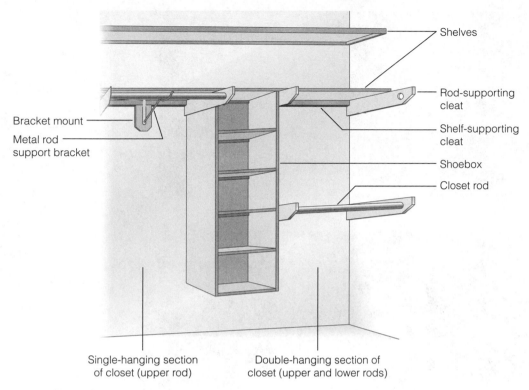

FIGURE 7-1 Components of a basic closet

TYPES OF CLOSETS

Closet designs range from simple to elaborate, depending on the customer's needs and desires. A simple closet consists of a single shelf and one rod for hanging garments. Some closets are made up of a network of shelves and shoeboxes and have cabinets equipped with drawers for storage and rods for hanging items.

Cleats are fastened to the wall and give closet shelves support. **Rod-supporting cleats** are fastened onto sidewalls and support both the shelf and the closet rod. Metal **closet rod supports** reinforce the middle of the rod and shelf on long spans. Shoeboxes vary in size and design and are installed for general storage purposes.

The differences between a simple closet and an elaborate closet are that an elaborate closet will have

- More shelves

- Stiles and rails added to the shelves and shoeboxes

- More shoeboxes

- Shaped edges accentuating certain areas

- Cabinets with drawers for extra storage space

In addition, there are differences in the types of materials used for simple closets and for more complex designs. Efficient use of space is always an important consideration in designing a closet.

MATERIALS

Materials used for closet construction vary greatly, ranging from paintable composite materials to stain-grade woods of all species (**Figure 7-2**). The type of material used is determined by the customer's desires or wishes. Considering the closet's function, composite materials are now being widely used, one of which is MDF (medium-density fiberboard). The MDF material is cost-effective and looks excellent when caulked and painted. **Melamine** is another material that is widely used in closet and cabinet construction. It is similar to MDF and plywood and particle board; the only difference is that both sides of the material are covered with melamine, which is a tough plastic material that is able to withstand marring and minor impacts. Normally, ¾-inch material is used in closet construction; anything less than ¾ inch would not provide enough strength or stability.

FIGURE 7-2   Materials commonly used for closet construction: (a) MDF (b) Pine (c) Plywood (d) Cedar

Some people use raw cedar in their closets for its aromatic qualities. Plywood is used in some cases, with 1-inch by 2-inch stiles fastened flush with the surface to hide the layers of veneer. More extravagant designs tend to maintain architectural flow by using the same stain-quality wood used throughout the rest of the house.

Some people prefer the simplicity of wire closet systems (e.g., Rubbermaid wire closet systems and storage bins). The same basic principles for closet layout can be used when installing a wire closet system.

Closets should be built with the same attention to detail as the rest of the house. The finish carpenter must have the ability to complete any closet design using any type of material.

PROCEDURES: CONSTRUCT A CLOSET

SAFETY TIPS

- Pay strict attention when operating any tool.
- Make sure safety features are intact and functioning properly.
- Keep blades sharp on equipment.
- Always unplug tools before changing blades.

7-1 CONSTRUCTING THE CLOSET IN CORRECT ORDER

There is a logical order to the steps involved in building a closet system. Following that order prevents other tasks from being road-blocked. For example, the shoebox needs to be built and installed so that rod-supporting cleats can be attached to it. You would not want to install the shelf-supporting cleats without the shoebox in place, as they would interfere with one another.

The following is an acceptable order in which to construct a closet system:

1. Mark closet layout lines.
2. Build and install shoeboxes.
3. Install cleats (both shelf- and rod-supporting cleats).
4. Cut and install shelves (and add stiles and rails if called for).
5. Add metal rod support brackets (center supports for rods, beneath the shelves).
6. Install closet rods.

7-2 MARKING LAYOUT LINES

Plumb lines and level lines are necessary when constructing a functional closet. By placing lines on the wall in accordance with a plan or design idea, what we are doing is creating an actual-sized pattern on the wall. Layout lines ensure that shelves will be level and spaced equally and that shoeboxes will be plumb. Use a standard 2-foot or 4-foot level, or a laser level, to mark layout lines. A framing square is sometimes helpful in continuing the layout lines on sidewalls if there is not enough room to use a level, as shown in **Figure 7-3**.

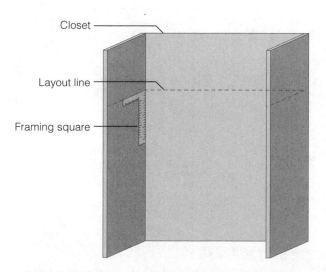

FIGURE 7-3 Using a framing square to scribe lines onto sidewalls

(a)

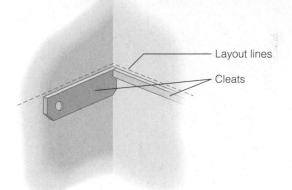

(b)

FIGURE 7-4 Cleats are nailed on layout lines

Layout lines determine locations where cleats will be nailed (**Figure 7-4**). A plumb line will help to ensure that shoeboxes are properly installed (**Figure 7-5**). Typical shelf and closet rod heights are described in the following section on determining shelf and closet rod heights. Following are the procedures for marking layout lines.

Measure from the floor and mark vertical shelving heights. Shelves normally begin 16 to 24 inches off the floor. Remaining shelves are spaced equally, ranging anywhere from 12 to 18 inches apart (**Figure 7-6**). In certain closets, rod heights determine the shelf heights.

Determine the location of the shoeboxes and mark the wall where sides will be. Once all shelf and shoebox locations have been measured and marked, use a 2- or 4-foot level, or a laser level, to make continuous lines from wall to wall and plumb lines representing the sides of the shoeboxes (**Figure 7-7**). Use a framing square to continue lines onto sidewalls if there is not enough room to use a level.

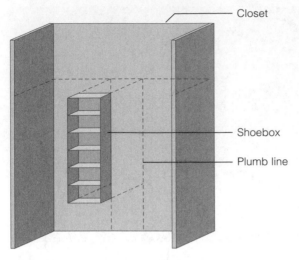

FIGURE 7-5 A plumb line ensures a properly installed shoebox

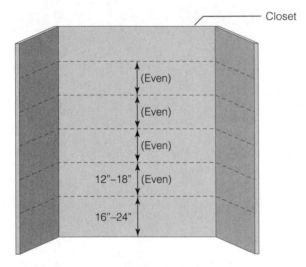

FIGURE 7-6 Spacing of shelves

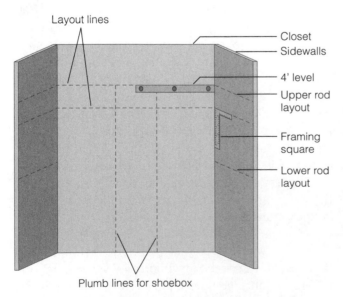

FIGURE 7-7 Diagram of layout lines in a closet

NOTE

Floors are not always true, so use a level of some type when making continuous lines from wall to wall. ∎

Determining Shelf and Closet Rod Heights

To mark layout lines in a closet, you must be familiar with typical shelf heights. In a closet designed for something such as linen or miscellaneous storage, the shelves begin anywhere from 16 to 24 inches off the floor. The remaining shelves are spaced anywhere from 12 to 18 inches apart, as shown in **Figure 7-8**. In a closet designed for hanging clothes, the rod height determines the shelf height, as the shelf sits on rod-supporting cleats.

The term **double-hanging** refers to a section of closet having both an upper and lower closet rod. The term **single-hanging** refers to a section of closet having only one rod. Single-hanging setups are preferred in some cases so that longer garments may be hung without interference from a lower rod. There are also **triple-hanging** sections: In closets with 9-foot ceilings, it is possible to install three closet rods, one at 3 feet, one at 6 feet, and one 2 or 3 inches below the 9-foot ceiling (the top rod is usually used for storing out-of-season clothes).

Figure 7-1 (at the beginning of the chapter) shows a closet with both single-hanging and double-hanging rods. It is common for a closet to have a combination

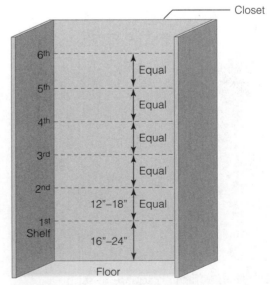

FIGURE 7-8 Shelf spacing in a typical closet layout

of both single- and double-hanging rods. Rods for hanging garments are typically spaced as follows:

- Lower rod height: 33 to 36 inches off the floor
- Upper rod height: 66 to 72 inches off the floor

> **NOTE**
>
> Since most closets do not have windows, a halogen work light is often useful when constructing a closet if there is no functioning light in place. ∎

7-3 BUILDING AND INSTALLING SHOEBOXES

Shoeboxes, which are used for both shoes and general storage, can vary in dimension. Vertical spacing of shelves is usually laid out equally from top to bottom. The width of shoeboxes commonly ranges anywhere from 12 inches to 48 inches, and more in some cases. The depth of a shoebox commonly ranges anywhere from 12 inches to 24 inches. The height of the shoebox usually corresponds with shelf heights, as previously discussed, so that the shoebox and the shelves can tie in together (see Figure 7-1). The most commonly used dimensions for a shoebox are as follows:

- Width: 12 to 48 inches
- Depth: 12 to 24 inches

Options for Shoeboxes

Many options are available for shoeboxes. Following are some of the most common ones:

- Adjustable shelves (**Figure 7-9**)
- Dadoes cut for shelves (**Figure 7-10**)
- Stiles and rails on shelves and shoeboxes (**Figure 7-11**)
- Shoeboxes that rest on the floor (**Figure 7-12**)
- Suspended shoeboxes (those that do not rest on the floor; **Figure 7-13**)

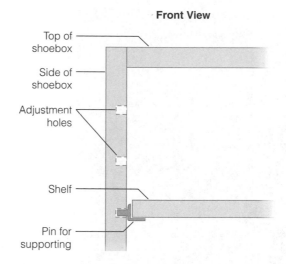

FIGURE 7-9 Adjustable shelves

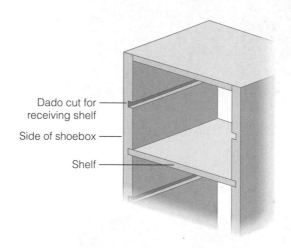

FIGURE 7-10 Dadoes cut for shelves

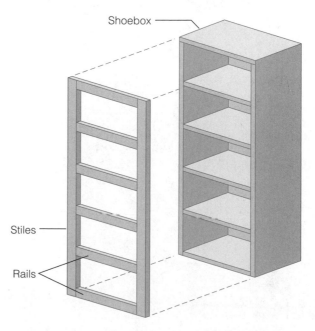

FIGURE 7-11 Stiles and rails on front of shoebox

FIGURE 7-12   A shoebox that rests on the floor

FIGURE 7-13   A suspended shoebox

Building Shoeboxes

Shoeboxes need to be built in logical order. An acceptable order for building a shoebox is as follows:

1. Rip sides and shelves on a table saw.
2. Cut sides to length, and then sand or plane off saw marks.
3. Cut shelves to length and sand or plane off saw marks.
4. Shape edges with router (if called for).
5. Mark sides with shelf layout lines. (A framing square is ideal for this.)
6. Assemble shoeboxes using finish nails and wood glue.
7. Nail cleats to back of shoeboxes, beneath the shelves.
8. Install shoeboxes.

The following text covers these procedures for building shoeboxes in more detail.

Rip the shoebox sides and shelves on the table saw. Remove saw marks from edges that are going to be exposed. A palm or orbital sander is ideal for this task. A hand plane can also be used to remove saw marks from the edges.

> **NOTE**
>
> MDF is available in a 1-inch by 12-inch size, a dimension that is ideal for closet components (e.g., shoeboxes and shelves). ∎

> **CAUTION**
>
> **Properly adjust blade depths (usually ⅛ to ¼ inch more than the thickness of the stock being cut).**

Once the sides have been cut, mark shelf layouts using either a framing square or a speed square (**Figure 7-14**). Space layout lines accordingly and evenly (see the section on determining vertical layout for shoeboxes). Some carpenters mark two lines and let the shelves rest between them during assembly. It is helpful to also make a line on the other side of the shoebox for a reference when nailing. Trying to nail without a reference line sometimes results in nails missing the shelf, and removing missed nails damages the surface of the material, creating more work for the

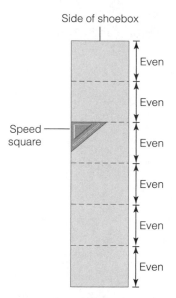

Side of shoebox

Even

Even

Speed
square

Even

Even

Even

Even

FIGURE 7-14   Mark shelf layouts onto sides of shoebox with a framing square or speed square

NOTE

If working with material that is going to be stained, remember that pencil lines will need to be sanded away completely. Mark material lightly, marking at the front and back instead of marking a continuous line that runs all the way across. Otherwise the pencil lines will be visible on the finished product. Sand or plane away saw marks before getting too far into the project. ∎

paint contractor and possibly affecting the look of the finished product.

Before cutting the shelves of the shoebox, take into account the added thickness of the sides. If the total width of the shoebox is supposed to be 24 inches, cut the shelves 1½ inches less, which would be 22½ inches. If there are going to be adjustable shelves, you may even have to cut them less than that, depending on the type of adjustable hardware being used.

Figure 7-15 shows a couple of different types of adjustable shelf hardware. If the shoebox is going to have a shelf over it, going from wall to wall (see Figure 7-1), there is no need to cut a 22½-inch shelf for the top. If any routing is to be done, do it before nailing the shoebox together.

Determining Vertical Layout for Shoeboxes

A common height for a shoebox is 84 inches (7 feet) off the floor. However, since the sides of the shoebox rest on top of the base molding, they will need to be cut less than 84 inches. To get the length for the shoebox sides, subtract the height of the base molding (including the distance the base is shimmed up off the floor) from 84 inches. An equation for calculating the length of the shoebox side is as follows:

84" (desired finished height of shoebox)
    − 3½" (height of base)
    − ⁷⁄₁₆" (distance base is shimmed off floor)

An easy way to mark the height as well as shelf layout lines onto the shoebox side is by using a

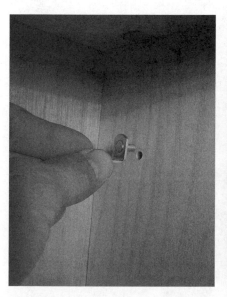

FIGURE 7-15   Adjustable shelf hardware

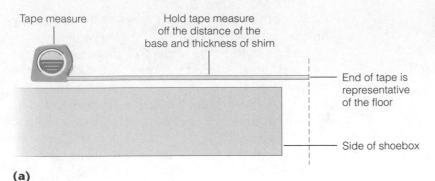

FIGURE 7-16 Placement of tape measure when cutting an inch

- Tape measure
- Hold tape measure off the distance of the base and thickness of shim
- End of tape is representative of the floor
- Side of shoebox

(a)

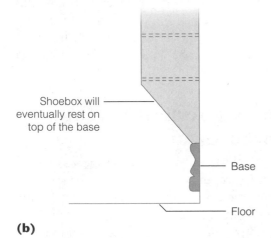

- Shoebox will eventually rest on top of the base
- Base
- Floor

(b)

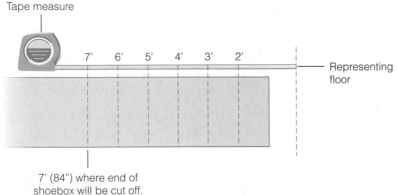

FIGURE 7-17 Marking the layout when cutting an inch

- Tape measure
- 7' 6' 5' 4' 3' 2'
- Representing floor
- 7' (84") where end of shoebox will be cut off.

method similar to *cutting an inch,* except in this case you cut 3 and $^{15}\!/_{16}$ inches (the height of the base + the distance it is shimmed off the floor). Let the tape measure hang off the end of the shoebox side a distance of $3^{15}\!/_{16}$ inches, as shown in **Figure 7-16**. (Use a sample piece of the base to simplify things.) Now that the end of the tape is in a position representative of the floor, you can mark the shelf layouts.

Start the first layout at 2 feet, and then mark every foot after that (**Figure 7-17**). The 84-inch mark is where the side will need to be cut. These shelf layouts will correspond well with the recommended closet rod heights, because there is one shelf at 3 feet and another one at 6 feet. (See the section on determining shelf and closet rod heights.)

Use a framing square or speed square to continue the layout line across the sides of the shoebox. After you've made the layout lines, the design for the bottom can be marked and cut (**Figure 7-18**).

> **NOTE**
>
> After you've made one side for the shoebox, you can use it as a pattern for marking all other sidepieces, saving time and ensuring accuracy. On average, there are between four and eight shoeboxes for the closets in a new home. ∎

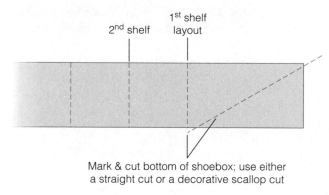

FIGURE 7-18   Marking and cutting shoebox bottom

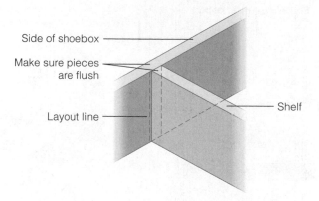

FIGURE 7-20   Make sure top edge of shelf is flush with side and shelf is resting between layout lines before nailing.

Assembling Shoeboxes

The worktable is the ideal location for assembling shoeboxes. The worktable brings everything up to a comfortable height. Position the sides and shelves as shown in **Figure 7-19**. Apply a thin ribbon of wood glue between the layout lines for each shelf. Using too much glue will result in squeeze-out and glue overrun. Be sure to remove any excess glue; this can be done using a cloth dampened with water. Glue squeeze-out and overrun will penetrate the wood and will not allow for proper staining.

Make sure the top edge of the shelf is flush with the side and the shelf is resting perfectly between the layout lines before nailing (**Figure 7-20**). Nail both sides of the shoebox to the shelves. To complete the shoebox, nail cleats beneath the shelves at the back of the box (**Figure 7-21**).

Installing Shoeboxes

To mount the shoebox, begin by locating wall studs. Mark corresponding stud locations onto the shoebox cleats. Predrill the cleats at stud locations using a

countersinking bit (**Figure 7-22**). It is easier to use a bit that drills both the pilot and countersinks. Hold the shoebox against the wall using the plumb layout line as a reference. Using 3-inch screws, screw through the cleats and into the wall stud (**Figure 7-23**). Once the shoebox has been installed, rod- and shelf-supporting cleats can be cut and fastened into place.

7-4 INSTALLING SHELF- AND ROD-SUPPORTING CLEATS

Cleats help secure shelves and shoeboxes to the wall. Cleats are fastened to wall studs with finish nails or 3-inch screws. The shelves are attached to the cleats with screws or finish nails (**Figure 7-24**). Cleats may vary in dimension. The typical dimension for a shelf-supporting cleat is 1¼ inches by ¾ inch. The ¾-inch

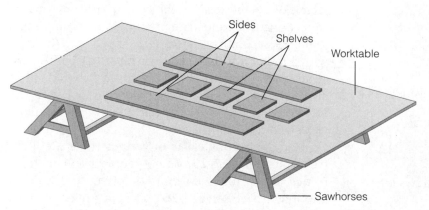

FIGURE 7-19   Positioning shoebox components onto a worktable

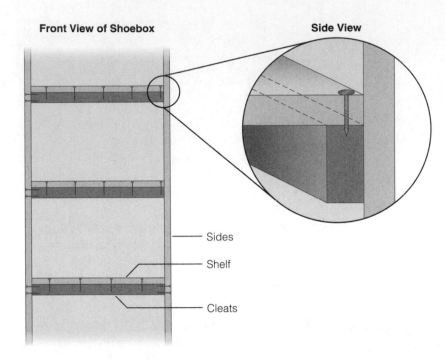

Front View of Shoebox

Side View

- Sides
- Shelf
- Cleats

FIGURE 7-21   Fasten cleats to the back of the shoebox, beneath the shelves.

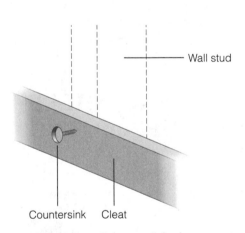

- Wall stud
- Countersink
- Cleat

FIGURE 7-22   Countersink cleats at stud locations.

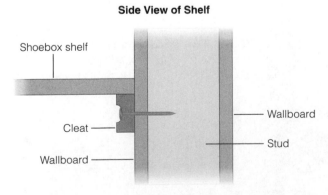

Side View of Shelf

- Shoebox shelf
- Cleat
- Wallboard
- Wallboard
- Stud

FIGURE 7-23   Mount cleats with 3-inch screws into wall studs.

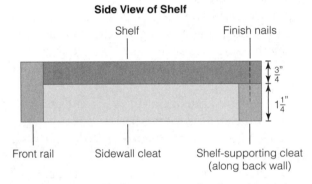

Side View of Shelf

- Shelf
- Finish nails
- $\frac{3}{4}$"
- $1\frac{1}{4}$"
- Front rail
- Sidewall cleat
- Shelf-supporting cleat (along back wall)

FIGURE 7-24   Shelves are attached to cleats with finish nails.

shelf plus the 1¼-inch cleat equal 2 inches, which is the exact dimension of the rail on the front of some shelves.

Cleats are made by ripping 1x material on the table saw to the desired dimension. The saw marks can be planed, sanded, or removed by truing on a jointer. A decorative edge can be shaped onto the cleat with a router if it flows with the closet design. All the cleats should match, meaning that they should either all be square or they should all be routed with the same design. You wouldn't want to have some square cleats and some routed cleats.

Rod-supporting cleats of various styles are illustrated in **Figure 7-25**. Some people use 1-inch by 6-inch cleats, to which brass or plastic rod holders are attached (**Figure 7-26**). Others prefer to have

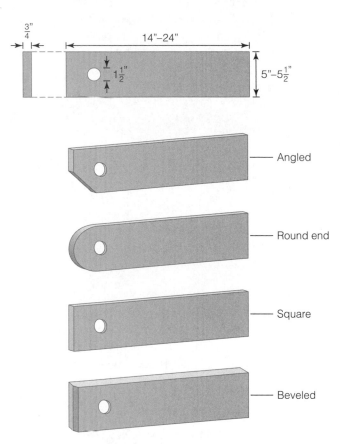

FIGURE 7-25 Different styles of rod-supporting cleats

(a)

(b)

FIGURE 7-26 Plastic rod holders attached to a 1-inch by 6-inch cleat

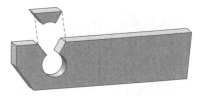

FIGURE 7-27 Keyhole slot cut into cleat enables the rod to be inserted and removed

FIGURE 7-28 Metal rod support bracket

the cleat itself supporting the rod; this is achieved by drilling a hole into the cleat that is slightly larger than the rod being used. Rods are usually 1¼ inches in diameter. A keyhole slot cut into one cleat enables a closet rod to be inserted and removed (**Figure 7-27**).

If the distance between rod-supporting cleats is greater than 4 feet, an additional means of support is necessary at or near the center. A metal rod support bracket provides additional support in the center of a closet rod (**Figure 7-28**). Metal rod support brackets are mounted onto a decorative block that is the same thickness as the cleat (**Figure 7-29**). Finished cleats are installed according to layout, which gives shelving something to rest on.

Making Rod-Supporting Cleats

Often the cleats are relatively far apart, which means that an additional means of support is necessary to reinforce the center of the closet rod. A closet rod longer than 4 feet needs additional support in the center. A rod support bracket is used to give extra support to

Side View of Shelf

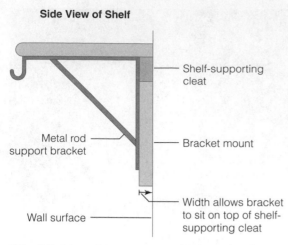

Shelf-supporting cleat

Metal rod support bracket

Bracket mount

Wall surface

Width allows bracket to sit on top of shelf-supporting cleat

FIGURE 7-29   Mount onto which a metal bracket is installed is the same thickness as the cleat

the center of longer closet rods. Support brackets differ in dimension, which means that the intended bracket will need to be used to figure out exactly where to drill the hole into the cleats so that the cleats and the brackets are matching. (The same method is used to determine the exact location of brass or plastic rod holders that are screwed onto the cleat.)

Figure 7-30 shows how to determine where holes need to be drilled into the cleats. If brass or plastic rod holders are being used, it is not necessary to drill holes

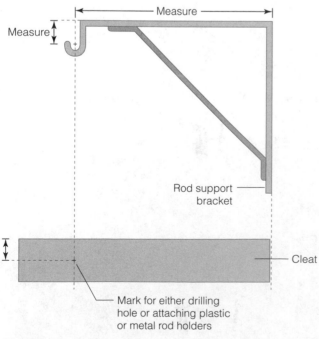

Measure

Measure

Rod support bracket

Cleat

Mark for either drilling hole or attaching plastic or metal rod holders

FIGURE 7-30   Determine where to drill hole into cleat

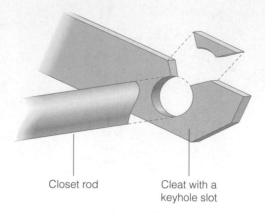

Closet rod

Cleat with a keyhole slot

FIGURE 7-31   Rod can be inserted and removed from a cleat that has a keyhole slot

into the cleats because these are simply attached with small wood screws. Once the hole positions are determined, use a 1½-inch spade bit to drill through the cleats. A 1½-inch spade bit also can be used to make wooden brackets that can be installed onto the cleat after the closet is complete. One bracket can be closed in, while the other one needs a **keyhole slot** cut into it. A keyhole slot will allow the rod to slide down into place later on (**Figure 7-31**).

> **NOTE**
>
> Spade bits cause damage as they exit through the back of material, so use a backing board (scrap material) or drill from both sides to avoid damaging materials. ∎

Once both rod-supporting cleats and shelf-supporting cleats have been made, you can nail them into place (**Figure 7-32**). Construction adhesive or wood glue can be used when attaching rod-supporting

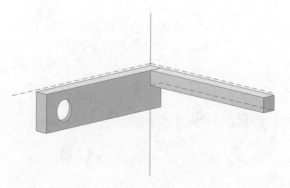

FIGURE 7-32   Nailing rod-supporting cleats into place

FIGURE 7-33 Taking measurement for shelf-supporting cleat

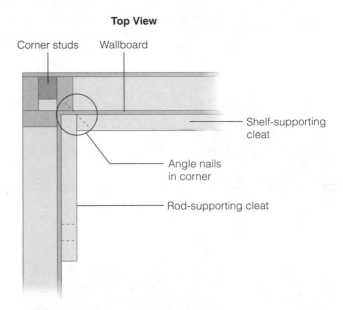

FIGURE 7-34 Angle nails in corners so that they do not miss wall studs.

cleats. When you are nailing cleats onto the side of a shoebox, you must use a shorter nail, along with wood glue. The combined thickness of the cleat and the side of the shoebox is 1½ inches. A nail shorter than 1½ inches must be used to avoid protruding finish nails.

After all side cleats have been installed, cleats along the back wall can be measured and cut. Take inside measurements for all remaining cleats (**Figure 7-33**). When nailing, keep cleats on level lines and make sure finish nails are going into studs. Nail cleats at every stud location. In corners, it may be necessary to slightly angle the nails in order to hit a stud (**Figure 7-34**). For shelves that do not have a closet rod beneath them, cleats of the same dimension can be used for the back and the sides.

> **NOTE**
>
> Use construction adhesive or screws to create a stronger bond when installing rod-supporting cleats. ∎

7-5 CUTTING AND INSTALLING SHELVES

Measurements for shelves may differ a bit from one shelf to the next. If the closet is terribly out of square, the ends of the shelves may have to be cut out of square so that a tight fit against the wall is achieved. To see if a square cut is going to work, use a framing square to check the wall as illustrated in **Figure 7-35**. Check the square at the 11¼-inch mark to find out how much adjustment is necessary. If a ¼-inch gap exists, the shelf will need to be angled accordingly. You can do this simply by making a slight adjustment corresponding with the out-of-square wall, as shown in **Figure 7-36**.

Measure and cut all the shelves, making adjustments if necessary. If shaping the edges with a router, do so before mounting shelves. Fasten shelves to cleats using finish nails. Remove any nails that missed the cleats. Shelves should be level and should rest upon securely attached cleats.

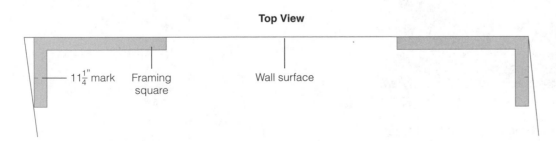

FIGURE 7-35 Use a framing square to determine how much adjustment is needed to properly fit shelf

Top View of Shelf

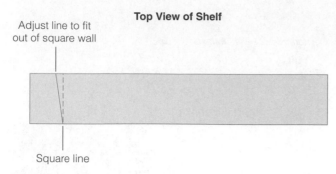

FIGURE 7-36 Use the framing square to scribe the shelf so that it corresponds to the out-of-square corner.

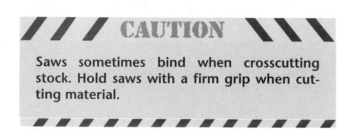

CAUTION

Saws sometimes bind when crosscutting stock. Hold saws with a firm grip when cutting material.

Wraparound Shelves

Some closets are designed with wraparound shelves. All steps previously covered apply to wraparound shelves as well, with the addition of one more step. Wraparound shelves join one another in the corners (**Figure 7-37**). To reinforce joining shelves, attach a **scab** to the underneath side. A scab is installed over

Side View

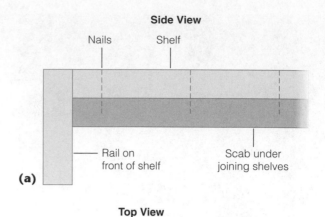

(a)

Top View

(b)

FIGURE 7-38 A rail is sometimes used to hide the scab beneath joining shelves

a joint to give it strength. Scabs should be glued and can be attached with either nails or screws. Usually a rail is added to the front of the shelf in order to hide the scab (**Figure 7-38**). An alternative to using a scab is to use a metal **gusset**. A gusset serves the same purpose as a scab; it is installed over a joint to give it strength. Square or rectangular gusset plates with predrilled holes can be found at any building supply or hardware store. Secure gussets with screws (make sure the screws are less than ¾ inch in length

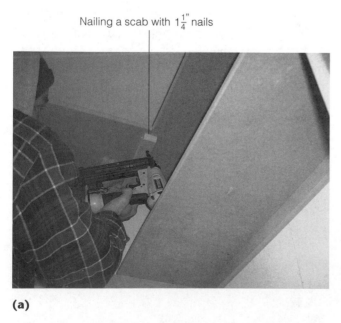

(a)

FIGURE 7-37 Wraparound shelves join in the corners

Top View

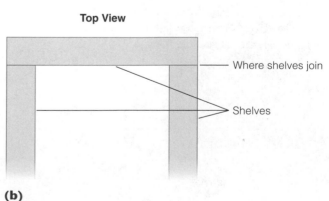

(b)

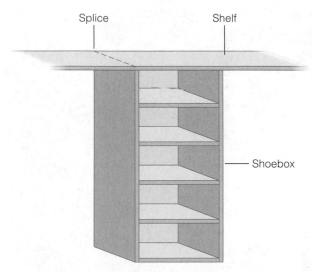

FIGURE 7-39 Splices can occur over one side of a shoebox.

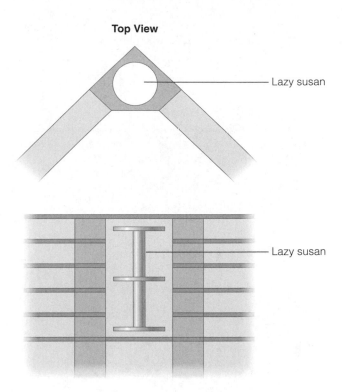

FIGURE 7-40 Lazy susan in a pantry

so that they do not go completely through the shelf material). Gussets are not as bulky as scabs and therefore are less noticeable. A good place for a splice to occur, if the shelf is not long enough to go from wall to wall, is over one side of a shoebox (**Figure 7-39**).

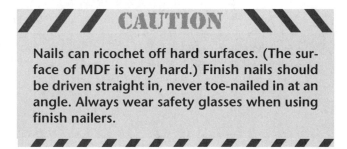

CAUTION

Nails can ricochet off hard surfaces. (The surface of MDF is very hard.) Finish nails should be driven straight in, never toe-nailed in at an angle. Always wear safety glasses when using finish nailers.

Pantry Shelves

Pantries are normally located in or near the kitchen. Typically, canned goods and food items that do not require refrigeration are stored in pantries. The components of a pantry are the same as those of a regular closet. The only difference may be in the depth and vertical spacing of the shelves, which must accommodate canned goods and smaller items. Smaller shelves are a more efficient use of space. Much space would be wasted if all the shelves were spaced 16 inches apart vertically throughout the entire pantry. Some pantries are equipped with a

circular rotating shelf system called a lazy susan, which saves space and makes working in the kitchen more efficient (**Figure 7-40**).

7-6 ADDING METAL ROD SUPPORT BRACKETS

Closet rods and shelves need additional support if their span is longer than 4 feet (**Figure 7-41**). **Figure 7-42** provides examples of different styles of bracket mounts, which give metal rod support brackets something to mount to. The supporting unit needs to be fastened to the wall, preferably at a stud location. The top of the bracket is screwed to the underside of the shelf. The bracket also provides extra support for the shelf, preventing the shelf from possibly sagging in the future.

Upper shelves need support also. Keep all shelf supports in line with one another so that everything appears uniform and well planned. **Figure 7-43** shows a support for an upper shelf. (Notice how it sits directly above the metal rod support.)

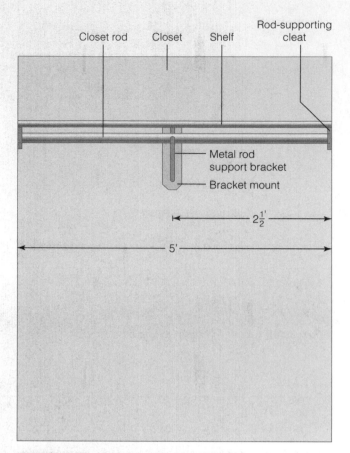

Closet rod   Closet   Shelf   Rod-supporting cleat

Metal rod support bracket

Bracket mount

$2\frac{1}{2}'$

5'

FIGURE 7-41 Longer rods need additional support in the center.

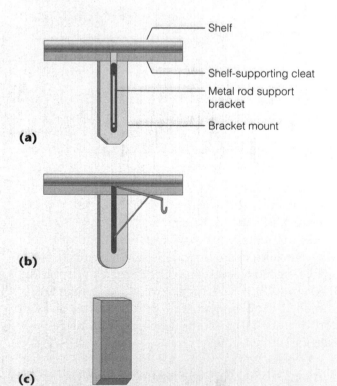

Shelf

Shelf-supporting cleat

Metal rod support bracket

Bracket mount

(a)

(b)

(c)

FIGURE 7-42 Different styles of bracket mounts

(a)

Shelf

Shelf-supporting cleat

MDF material cut at 45° angle

Bracket mount

Nail and use wood glue to create a strong support

(b)

FIGURE 7-43 Support for upper shelf sits directly above metal rod support

SUMMARY

- Closets range from simple to extravagant.
- Closets should be built with the same attention to detail as the rest of the house.
- Efficient use of space is a must when designing a closet.
- Normally, ¾-inch material is used for closet construction.
- Level layout lines make building closets easier.
- Shoeboxes can sit on the floor or can be suspended up off the floor.

- Shoeboxes need to be firmly attached; screws should hit the studs.
- Hanging space should not be cut short by an out-of-place shoebox.
- Cleats are nailed to the wall, whereas the shelves rest on the cleats.
- Any shelf or closet rod that is longer than 4 feet should get a support in the middle.
- Shelves should be level.
- Shelves should rest upon securely attached cleats.

KEY TERMS

cleats 1-inch by 2-inch pieces of trim that support shelves. The bottom of the cleats can either be left square or they can be rounded by using a router.

closet rod supports Metal brackets that support closet rods and shelves; installed at or near the center of a long closet rod to give support so that it does not sag.

double-hanging A section of closet having two closet rods, an upper and a lower rod.

gusset A thin piece of wood or metal that is installed over a joint to add strength where shelves join. Metal gussets are typically installed with screws.

keyhole slot A keyhole-shaped cut into a rod-supporting cleat that enables the closet rod to be inserted or removed.

melamine Building material that consists of a tough resin material, melamine, bonded to either side of

plywood, MDF, or particle board, creating a solid sheet of usable material. Melamine withstands marring and minor impacts. It is used widely in commercial jobs; one of its benefits is that it does not need to be painted.

rod-supporting cleats Cleats that also support closet rods; sometimes drilled with a hole and keyhole slot.

scab Usually a 1-inch by 2-inch piece of material (wood or MDF) that is installed over a joint to give it strength. It is similar to a gusset; its dimension is the only difference.

single-hanging A section of closet having only one closet rod.

triple-hanging A section of closet having three closet rods; an upper, middle, and lower rod.

REVIEW QUESTIONS

1. What is a typical spacing for the shelves of a shoebox?

2. What is the typical depth of a shoebox?

3. What are some different options when building a shoebox?

4. Why is it a good idea to use level layout lines when building a closet?

5. Can cabinets be incorporated into a closet design?

6. What are the design differences between a clothes closet and a pantry?

7. Should a 6-foot closet rod get a center support?

8. Explain how a shoebox is mounted to the wall.

9. When wraparound shelves meet in a corner, what can be done to strengthen this joint?

10. Is it okay to leave saw marks on closet materials?

11. What is the purpose of cutting a keyhole slot in a rod-supporting cleat?

12. What is a cleat?

13. What length of screw is ideal for mounting a shoebox?

14. Is there a set number of shelves for shoeboxes?

15. If you are using a material such as plywood, a couple of things can be done to hide exposed edges. What are they?

16. Is plywood suitable for receiving a routed edge?

Crown Molding

OBJECTIVES

After studying this chapter, you should be able to

- Measure for crown molding
- Cut crown molding
- Cope and splice crown molding
- Properly install crown molding
- Combine crown molding with other moldings

INTRODUCTION

Crown molding, sometimes referred to as **cornice**, provides a transition between the walls and the ceiling (**Figure 8-1**). Unlike other moldings, crown sits against the wall and ceiling at an angle, a situation that requires **compound cuts** (i.e., cuts made in two directions or at two different angles). Measurements and cuts have to be precise, especially when working with high-quality wood that is going to be stained. ■

Ceiling

Support block

Crown

Wall

(a)

(b)

(c)

(d)

(e)

FIGURE 8-1 Crown molding

TYPES OF CROWN MOLDING

There are many different types of crown molding—different **milling** patterns, profiles, and different widths—but all are basically cut and installed with the same methods. Material is milled when it is sent through a machine to be planed, cut, or given a profile. A profile is an end or side view and shows the molding's contour, height, and thickness. The amount of skill needed to complete crown installation equals that of a cabinetmaker.

MATERIALS

Crown is available in many different sizes and profiles, ranging in lengths from 8 to 16 feet. Crown is milled from both softwood and hardwood, both of which can be stain-finished. For a paint finish, finger-jointed material can be used, as well as MDF, plastic, vinyl and PVC, and fiberglass-reinforced gypsum. Crown made of MDF usually comes preprimed. Finger-jointed material is not suitable for staining.

 **PROCEDURES: INSTALL CROWN MOLDING**

 SAFETY TIPS

- Make sure safety features on all tools are working properly.
- Keep tools clean so sawdust and other material will not prevent the safety features from working.
- Inspect trim to make sure there are no hidden nails or staples before making cuts.
- Pay strict attention when operating tools, and do not overwork machines by cutting at a faster rate than they can handle.
- Adjust blades to proper depth, stabilize material, build up the rpm before the blades make contact with stock, cut in a direction that is away from the body, and use backing boards and push sticks when needed.

8-1 MEASURING CROWN MOLDING

Finish carpenters develop their own personalized methods of performing their duties, as stated earlier. Although their methods may be different, all finish carpenters have the same common goal: to produce a finished trim job with exquisite quality, and to accomplish that in a safe manner.

As with base molding, when installing crown, some carpenters prefer to measure, cut, and install one piece at a time, and then repeat the process for the next piece. Other carpenters measure an entire room first, and then cut and install all the pieces. Still others take all the measurements for the crown throughout the house, and then cut the pieces, number them, and finally install them. Each method has certain advantages and disadvantages; ultimately, you must decide which one will work best for you in producing quality work in a safe manner.

Unlike base molding, crown needs to be measured and installed in a clockwise direction around the room, moving from left to right. The reason for this is that the physical act of coping (cutting a trim's profile in a way that allows it to fit precisely over an adjoining piece of trim) is made easier this way (at least for a right-handed person); the angles involved in cutting and coping crown are more easily achieved when working in a clockwise direction, especially when considering the necessary angle of the coping saw in relation to the angle of the piece being coped. (See the section later in this chapter on coping crown molding.)

> **NOTE**
>
> When measuring a room for crown, start at a strategic location, such as an *outside corner*, so that the final piece will not have to be coped at both ends. ∎

You want to avoid having to cope both ends of crown whenever possible because, as stated earlier, it is more difficult to cope one end of the crown because of the necessary angle of the saw blade in relation to the angle of the piece being coped. An outside corner

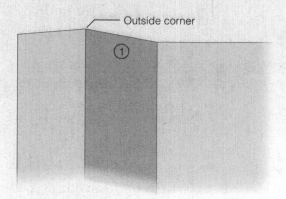

FIGURE 8-2 Outside corners are an ideal beginning point for crown.

is an ideal place to begin when installing crown (**Figure 8-2**). Some rooms are square or rectangular and have only four walls, which means the final section of crown will have to be coped at both ends. Coping both ends of a section of crown requires a little more skill and effort; therefore, it is best to begin at a well-thought-out point, keeping the finished piece in mind when doing so.

Crown measurements are often longer than base measurements, since there are no doorways to break up the wall space. There is no floor on which to rest the tape or block, so measurements are taken with a different method. You can either use two people, one holding each end of the tape measure, or simply position a ladder in the center of the wall being measured. Butt the end of the tape measure against the wall, pull it to the center of the wall (or close to it), and find an easy-to-remember mark (e.g., 6 feet). Make a pencil

mark there. Then butt the end of the tape against the other corner and pull it to the mark made previously (i.e., 6 feet). Add the two measurements together to find the total length (**Figure 8-3**). Write the total length on the wall, and later jot all the measurements down on paper and take them to the miter saw station. An entire room can be measured this way.

> **/// / CAUTION \ **
>
> **If using scaffolds, make sure to use safety railings as required by OSHA. Make sure wheels are locked so that scaffolding cannot shift or move unexpectedly. Keep ladders and scaffolds on level ground. Do not roll scaffolding over extension cords or air hoses, as this can cause scaffolding to tip over and can damage cords and air hoses.**

Figure 8-4 shows the same illustration used in Chapter 4 on base molding. The illustrated room shows typical angles and how the miter saw must be set to properly cut the angles. Crown molding is cut upside down on the miter saw, which is one reason why installing crown molding is considered a difficult and confusing task. Since crown is cut upside down, you may want to reverse the symbols while recording the measurements; reversing the symbols will refresh your memory and serve as a reminder later on at the miter saw, showing you exactly which end gets what type of angle. (See the section later in this chapter on cutting crown molding.)

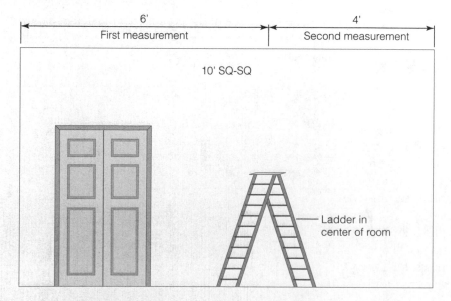

FIGURE 8-3 Measuring from the center of the wall

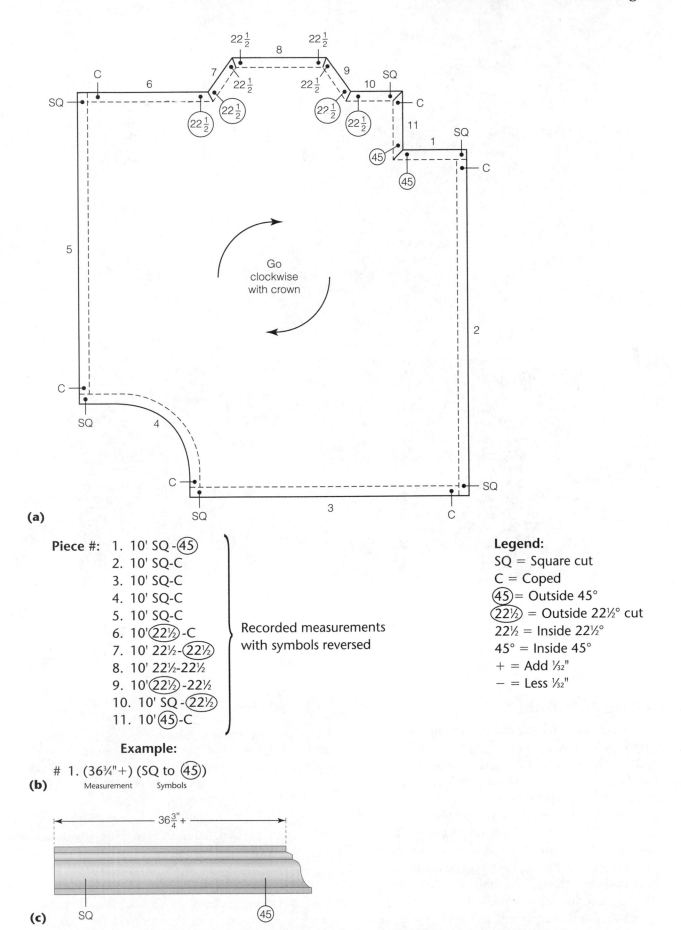

(a)

Piece #: 1. 10' SQ -(45)
2. 10' SQ-C
3. 10' SQ-C
4. 10' SQ-C
5. 10' SQ-C
6. 10'(22½)-C
7. 10' 22½-(22½)
8. 10' 22½-22½
9. 10'(22½)-22½
10. 10' SQ -(22½)
11. 10'(45)-C

Recorded measurements with symbols reversed

Legend:
SQ = Square cut
C = Coped
(45) = Outside 45°
(22½) = Outside 22½° cut
22½ = Inside 22½°
45° = Inside 45°
+ = Add ⅟₃₂"
− = Less ⅟₃₂"

Example:

1. (36¾"+) (SQ to (45))
(b) Measurement Symbols

(c) SQ (45)

36¾"+

FIGURE 8-4 Room diagram showing angles and symbols for recording measurements

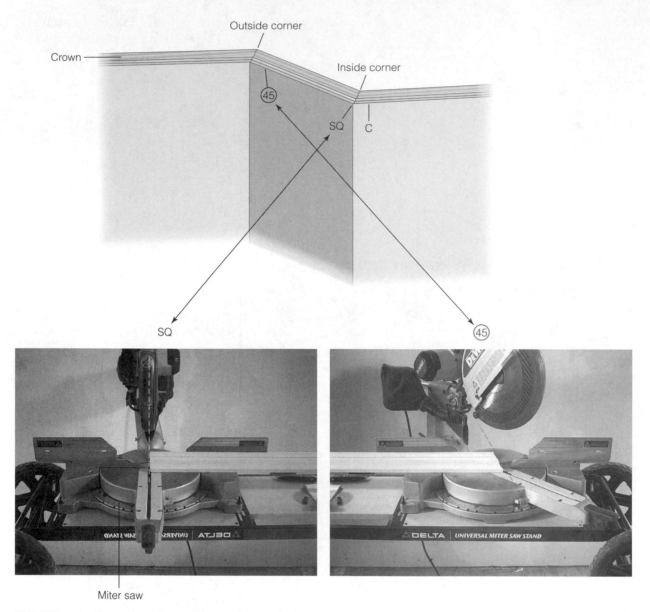

FIGURE 8-5 Reversing the symbols of recorded measurements

Reversing Symbols

Many people are discouraged from attempting to work with crown molding because of the difficulty of remembering the direction of the angles that are needed when the pieces of crown are sitting inverted on the saw base. The finish carpenter can lessen such confusion by reversing the order of the symbols when recording the measurements for crown. **Figure 8-5** shows an example of how symbols can be reversed so that when you are cutting the necessary angles at the miter saw, the degree symbols can be easily followed. By recording the symbols backward, you can more easily cut the angles at each end of a piece correctly.

> **NOTE**
>
> When you are cutting crown, the fence of the miter saw represents the wall, and the base of the saw represents the ceiling. (For more details, see the section later in this chapter on cutting crown molding.) ■

Keep the following guidelines in mind when measuring for crown molding:

- Work in a clockwise direction.

- Instead of trying to measure an entire wall section, take two measurements from the center of the wall, and then add them together to determine the total length.

- Begin at an outside corner so that both ends of the last piece do not have to be coped.

- Reverse the symbols when recording measurements for crown molding.

Determining Height of Crown

The height of the crown needs to be determined primarily so that

- Support blocks can be made.

- Crown can be properly cut.

- Crown can be properly installed.

Since crown molding comes in different sizes, it is important to establish the height of the particular crown being used.

To determine the height of the crown, cut a sample piece of the stock being used and position the piece as shown in **Figure 8-6**. Notice that the crown does not sit at a true 45-degree angle. When the crown is seated properly, as illustrated in Figure 8-6, make a mark at the bottom of the crown, and also mark the inside top and bottom of the piece (for getting the dimensions of the support blocks). These measurements will be used for making **support blocks**, and the bottom measurement will be used for the proper cutting and installation of the crown. Support blocks are triangular-shaped blocks that are installed in places where there is nothing to nail the crown to, such as on walls where there are no ceiling joists. Support blocks,

glued with construction adhesive and nailed securely to the framing (the top-plate), give something solid to nail the crown to (see the section *Support Blocks*).

Measure down at each corner of the room and mark the proper crown height so that chalk lines can be snapped (**Figure 8-7**), or cut a **gauge block** and use it to mark the corners and several locations between the corners of the room. These lines are where the bottom of the crown will rest when you install it. A gauge block can be cut from scrap material; it is used when there is a need to make many marks of the same length. Using a gauge block saves both time and energy and reduces the chance of making a mistake. Most finish carpenters prefer marking the crown height using a gauge block versus using chalk lines because the crown molding needs to follow the subtle imperfections found with most ceilings (meaning that ceilings usually are not perfectly flat). Make the same height mark on the fence

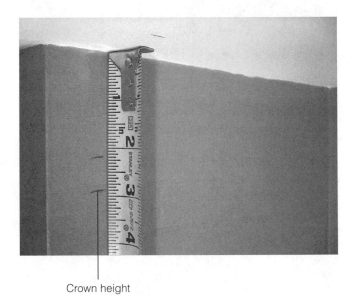

Crown height

(a)

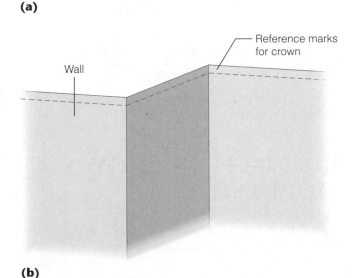

Reference marks for crown

Wall

(b)

FIGURE 8-7 Making reference marks along the ceiling

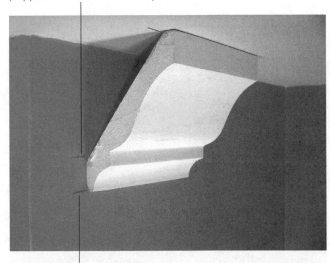

Mark inside top and bottom (support block measurements)

Mark bottom (crown height)

FIGURE 8-6 Positioning a sample of the crown being used

(a)

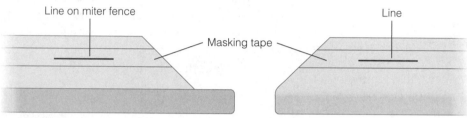

(b)

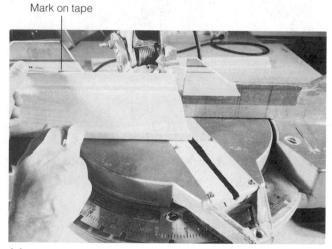

(c)

FIGURE 8-8 Making a reference mark on the fence of the miter saw

of the miter saw (see **Figure 8-8**), so that you can hold the crown on that mark while cutting it.

8-2 CUTTING CROWN MOLDING

Crown molding can be cut in several different ways. Most professional finish carpenters prefer the method discussed here for the practical reason that it is faster and simpler than any other method. The more complex methods of cutting crown allow for more mistakes,

especially when used by beginners. Once you have achieved a basic understanding of making miter cuts on crown molding, you can easily adopt other methods of cutting the crown if you so desire.

After determining the height of the crown, you can mark the height measurement onto the fence of the miter saw. Painter's masking tape, blue or white, can be placed on the fence of the saw to mark the height of the crown. The mark on the tape is where the crown will need to be held when it is being cut (see **Figure 8-9**). Take particular notice of the following:

- Crown sits upside down on the miter saw. The fence of the miter saw represents the wall, and the base represents the ceiling.

- The small cove represents the bottom of most crown moldings.

- Crown sits on the height mark scribed onto the miter fence.

- The height mark on the miter fence is the same as the height mark on the wall (which represents the bottom of the crown).

- Crown sits at the same angle while being cut and while being installed.

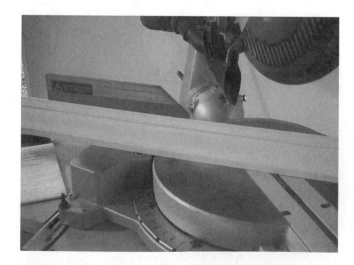

FIGURE 8-9 Crown being held against the reference mark on the miter fence

NOTE

In using this method, remember that the crown is *always* cut upside down. This never changes. Other methods require you to turn the crown in different directions, depending on the desired cut. Once you have mastered basic cuts, you can easily begin making compound miter cuts. ∎

CAUTION

Hold crown firmly while cutting. Keep hands and fingers a safe distance away from the blade.

Some people use **crown stops** to aid them in cutting crown. Crown stops are mounted onto the miter saw and they are designed to hold the crown at the correct height, or in the correct position, while it is being cut. Using crown stops has advantages and disadvantages. An advantage is that the crown stops hold the crown in the correct position so that accurate results are produced each time the trim is cut. A disadvantage to using crown stops is that when two or more people are using the same miter saw to cut two different types of trim (e.g., crown and base), the crown stops get in the way and are constantly being taken off and put back on. When crown stops are not used, the piece must be held firmly and squarely against the miter fence and directly on the height mark (**Figure 8-10**).

All measuring takes place at the bottom (near the small cove) on the back edge of the piece (the part that will actually make contact with the wall). Make a pencil mark on the thin edge of the bottom where you can see it while cutting (**Figure 8-11**). If possible, first cut whichever end will afterward allow you to hook the

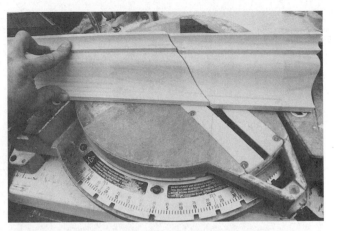

FIGURE 8-10 Hold crown firmly against the miter fence in line with the reference mark.

FIGURE 8-11 Mark crown along the edge where it can easily be seen while cutting.

tape and pull a measurement, so that you will not have to "cut an inch" to get the measurement. Sometimes cutting an inch is unavoidable, however. **Figure 8-12** illustrates how to measure by cutting an inch. Just remember to add the lost inch. (See Chapter 4 for more information on cutting an inch.)

Figure 8-13 shows basic miter saw cuts.

FIGURE 8-12 Cutting an inch

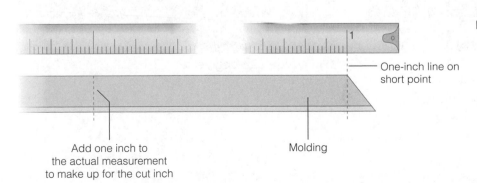

One-inch line on short point

Add one inch to the actual measurement to make up for the cut inch

Molding

(a) Square cut

(d) Inside 22½-degree cut

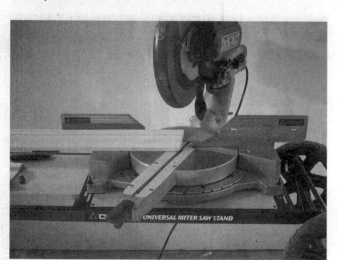

(b) Inside 22½-degree cut

(e) Inside 45-degree cut

(c) Inside 45-degree cut

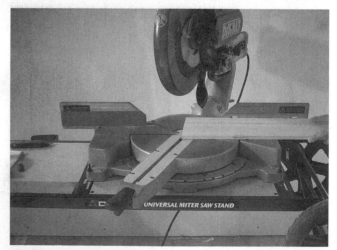

(f) Outside 22½-degree cut

FIGURE 8-13   Basic miter saw cuts

(g) Outside 45-degree cut

(h) Outside 22½-degree cut

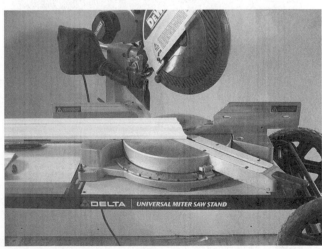

(i) Outside 45-degree cut

FIGURE 8-13 *(continued)*

8-3 COPING CROWN MOLDING

Trim will need to be coped where two pieces meet at a 90-degree inside corner. The first piece will go all the way into the corner and will be cut off squarely, whereas the adjacent piece will be coped to fit over the first piece (see **Figure 8-14**). Coped joints fit more tightly than mitered joints and are more likely to remain tight over time, despite any shifting or settling of the structure. The coped piece more or less locks the end of the first piece into position. Mitered inside joints have a tendency to open up from pressure caused by nailing, whereas a coped joint will not open.

Coping is done using a coping saw, after the piece of crown has received an inside 45-degree cut (**Figure 8-15**). Some prefer that the teeth of the blade angle back toward the handle of the coping saw, while others prefer to have the blade in the other direction. Since either way is acceptable, you may want to experiment a little and try it both ways to find which way works best

FIGURE 8-14 A coped piece of crown fitting over a square piece of crown

(a)

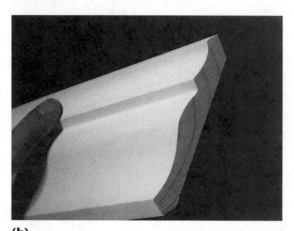

(b)

FIGURE 8-15   A piece of crown, with an inside 45-degree cut, ready for coping

for you. Go with whichever way feels most comfortable for you.

The blade of the coping saw is thin and narrow so that the tight and intricate profiles of trim can be cut around with precision. Before coping, it is helpful to trace the profile of the trim with the edge of a pencil (**Figure 8-16**). Doing this better defines the profile in a visual sense, allowing you to cope without straining to see where the profile begins and ends.

Figure 8-17 shows a piece of crown as it is being coped. Take note of the angle of the blade in relation to the piece of crown. Such angles, referred to as **undercutting**, are necessary when coping crown; anything less would not permit the coped end to join tight with the other piece. Looking at the back of a coped piece of crown provides a different perspective of the undercut (**Figure 8-18**).

Figure 8-19 further illustrates the necessary angles. Crown, unlike base molding, needs a severe undercut in order to be joined properly. Once you have finished coping the crown, smooth the edges with sandpaper (**Figure 8-20**). Make sure to remove any pencil marks.

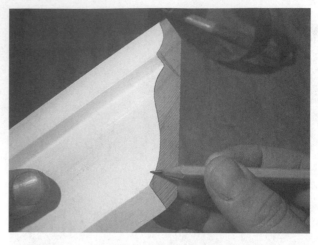

FIGURE 8-16   Trace the profile with the edge of a pencil.

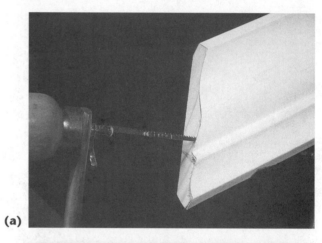

(a)

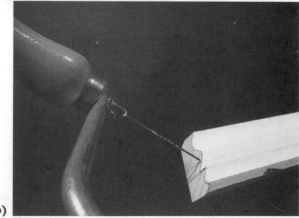

(b)

FIGURE 8-17   Coping crown in progress

> **NOTE**
>
> Measure and mark the piece of trim before coping, while you still have something to hook the tape onto. ■

FIGURE 8-18 The undercut as viewed from the back

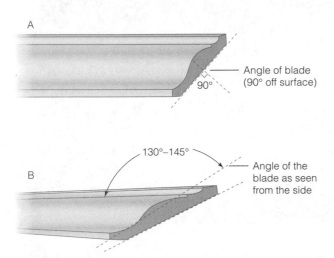

FIGURE 8-19 The angles needed to cope crown properly

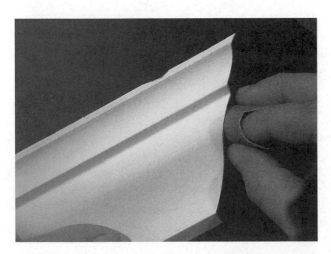

FIGURE 8-20 Fine-tune the edges of the cope using fine-grit sandpaper.

When coping crown, remember the following:

- Cope trim at inside 90-degree corners.

- Trace the trim profile with a pencil after making an inside 45-degree cut.

- After coping, smooth the edges with fine-grit sandpaper and remove pencil marks.

8-4 SPLICING CROWN MOLDING

Splices, which are also considered scarf joints, should consist of 45-degree cuts and should occur over ceiling joists or support blocks so that solid joints are

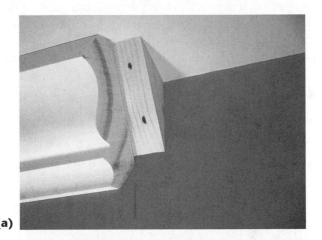

(a)

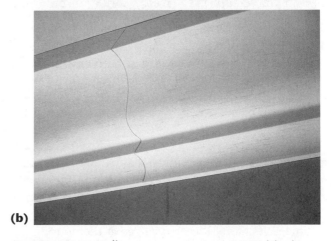

(b)

FIGURE 8-21 Splice crown over a support block.

created (**Figure 8-21**). Splices should be glued with wood glue, preferably a waterproof glue that will resist fluctuating humidity levels. After splicing, any slight imperfections should be sanded smooth so that the finished product is seamless. An imperfect joint will be highly visible on the finished product.

When using a dentil crown or a molding that has a repeating pattern (such as egg and dart molding), splice the two pieces so that the pattern is not interrupted. (See the sections later in this chapter on stacked and dentil crown molding.)

8-5 INSTALLING CROWN MOLDING

Crown molding can be installed once the room has been prepared (lines snapped, studs and joist marked, support blocks nailed in place, etc.). Start installation with the piece designated as number one (preferably at an outside corner, as mentioned in the earlier section on measuring crown molding). Continue installation of the crown in a clockwise direction, moving from one wall to the adjacent wall, fastening the molding with finish nails at joist and support block locations.

Following are some of the reasons why pieces of crown molding should be numbered and installed in a consecutive order:

- It is an efficient system, and the one that works best for maintaining consistency on a regular basis.

- It is less confusing, versus jumping around and skipping wall spaces (which involves a lot of backtracking and remeasuring).

- A coped piece of crown fits over the prior piece, so each piece *depends* on the prior piece being installed first.

- Skipping wall spaces means that many pieces of crown will have to be coped at both ends in order to fit. Coping crown at both ends is difficult to do and usually results in one end having a less-than-perfect joint.

Leave the ends of the crown loose so that you can make adjustments if necessary (**Figure 8-22**). In order to fit pieces of crown together perfectly at inside and outside corners, the pieces may need to be moved up or down slightly in order to close small gaps. Crown sits at an angle; therefore, if there is a gap at the bottom, the pieces need to be moved *down* together. If there is a gap at the top, they need to be moved *up* together in order to close the gap. If joined pieces of an outside corner gap due to an out-of-square corner, shave off part of the back of the pieces with a utility knife (**Figure 8-23**).

In order to create a perfect joint, sometimes it may be necessary to go slightly above or below the line representing the bottom of the crown. A slight

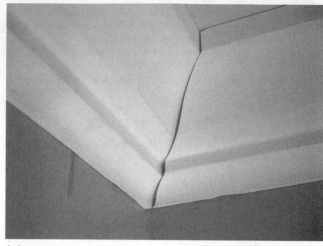

(a)

(b)

FIGURE 8-22 Leave the ends of crown loose so that adjustments can be made: (a) Pieces need to move down together to close the gap; (b) Pieces need to move up to close the gap.

variation is acceptable, especially if it is for the sake of creating a tighter joint. Once the pieces are joined perfectly, they can be fastened with finish nails.

If crown is not long enough to go from wall to wall, a splice, or scarf joint, may be needed (see the earlier section on splicing crown molding). Splices should occur over something solid. Use a support block where splices occur (refer to Figure 8-21).

If both ends of the crown have been coped, nail both ends securely, and then nail the middle. A piece of crown that is coped at both ends should be cut a fraction longer than necessary so that the middle is bowed out. Then, when the ends of the piece are nailed, the middle will be pushed into place and nailed, which creates a pressure lock and makes the joint at both ends tight.

A decorative block for the crown to tie into can be made or purchased, giving you another option for stopping the crown. (See **Figure 8-24**.)

(a)

(a)

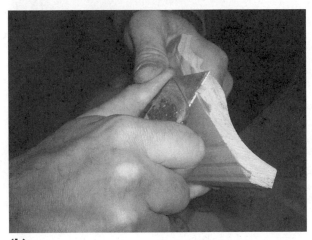

(b)

(b)

FIGURE 8-23 A gap at an outside corner that does not join properly can sometimes be solved by carving out the back.

Ending Crown Molding

Sometimes crown has to be terminated at a certain point, for example, when ceiling heights change drastically. There are several ways to end the crown in such situations. One way is to make a 45-degree cut so that it returns straight back into the wall. Another way is to cut the pieces so that they return to the wall at a 45-degree angle. (This method is similar to that used for apron return pieces when installing windows; see the section in Chapter 5 on installing window aprons.)

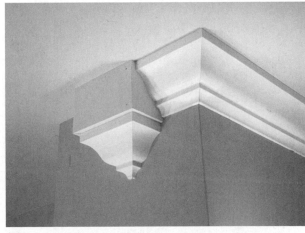

(c)

FIGURE 8-24 Options for terminating crown: (a) 45-degree return; (b) 90-degree return; (c) Decorative box.

SUPPORT BLOCKS

Support blocks are used in places where there are no ceiling **joists** to which to nail crown molding. Framing carpenters usually place **deadwood** (or **nailers**) in areas where there are no ceiling joists. The nailers provide something to nail to, but sometimes they do not extend out far enough for the crown to be fastened securely to them (**Figure 8-25**). **Figure 8-26** shows two situations, one where blocks are not needed, and one where they are needed.

The carpenter must determine whether or not blocks will be needed by locating hidden nailers (e.g., joists or deadwood). Use a finish nail or stud finder to locate nailers. To locate hidden nailers, start at the furthest point and work inward, toward the wall, until solid wood is found. If there is a need for support blocks, cut them to the appropriate size and nail them

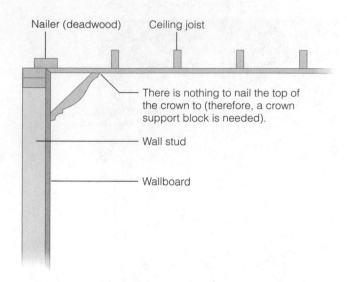

FIGURE 8-25   Framing nailers that do not extend far enough to fasten the crown to them

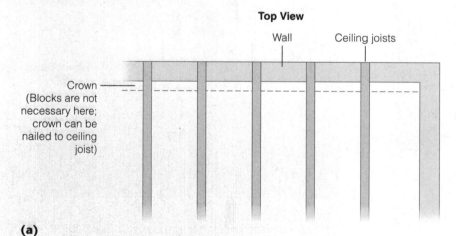

(a)

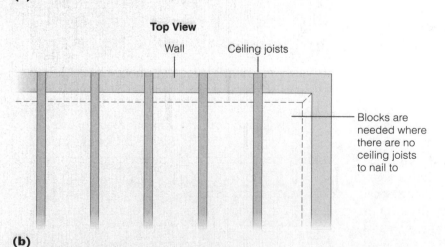

(b)

FIGURE 8-26   Diagrams showing where blocks are needed and where they are not needed

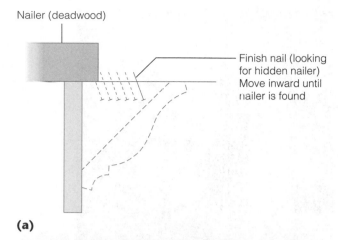

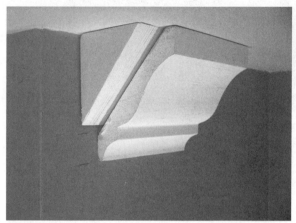

(b)

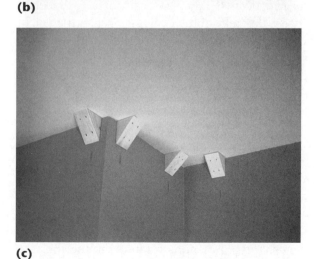

(c)

FIGURE 8-27   A support block nailed into place, showing proper grain pattern of block

as shown in **Figure 8-27**. Use adhesive on the surfaces of the block that make contact with the wall and ceiling. Figure 8-27 also shows the proper grain pattern for support blocks, which ensures a stronger block. Some carpenters prefer to cut blocks that are longer in length than what is actually needed so that they will have more nailing area. Also, it helps to cut

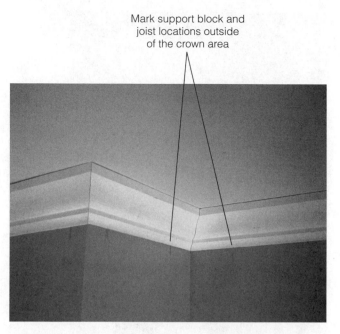

FIGURE 8-28   Make the marks indicating block locations outside of the crown area so the marks will be visible while installing the crown.

blocks in such a way that the corner of the block is clipped, as in Figure 8-27. Having a clipped corner will enable the block to seat better against the wall and ceiling (the reason for this is that joint compound creates a corner that is not square).

When using support blocks, place them at the corners and every 16 inches after that. On walls that have joists, locate the center of one with a nail; then all others should be measurable at every 16 inches. Place pencil marks indicating joist locations outside of the crown area so that the marks will not get covered up (**Figure 8-28**).

> **NOTE**
>
> Crown needs to be nailed to something solid. Wallboard is not strong enough and is not designed to securely hold nails or fasteners. ∎

STACKED CROWN MOLDING

All the methods discussed previously apply to stacked crown (see **Figure 8-29**, which shows stacked crown with the parts labeled). When incorporating a 1× material (i.e., material that is 1-inch normal thickness) into the trim scheme, simply **snap lines** around the room at the desired distance down from the ceiling. Snapping a

(a)

(b)

(c)

FIGURE 8-29   Stacked crown

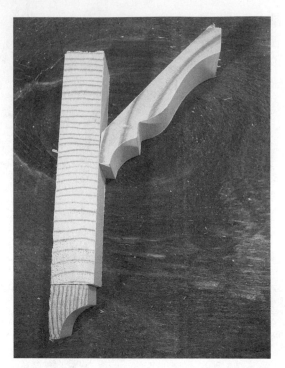

FIGURE 8-30   Arrange sample components of stacked crown.

line is done using a chalk line (filled with blue chalk—red chalk bleeds through paint). Make allowances for the cap at the bottom when figuring the total height of the stacked trim. If the plans call for a total height of 8 inches down from the ceiling, consider the height of the crown, the 1 × 6, and the bottom cap. If there is no predetermined distance, continue the theme or style that is present throughout the rest of the house.

You may wish to cut sample pieces of each stock and arrange them as shown in **Figure 8-30**, making a prototype of sorts. Arrange the pieces until the desired look is achieved, and then measure the point at which each piece will need to lie so that walls can be marked accordingly. Once that is done, you can snap lines for the 1 × 6 around the room. After the 1 × 6 has been installed, you can take measurements for the crown. The 1 × 6 does not affect the methods of taking measurements. Instead of hooking and butting the tape against the wallboard, hook and butt the tape against the 1 × 6. **Figure 8-31** shows the progression of installing stacked crown.

NOTE

Although support blocks for the top of the crown may still be needed, the 1 × 6 gives you something solid to nail the crown to at the bottom. ■

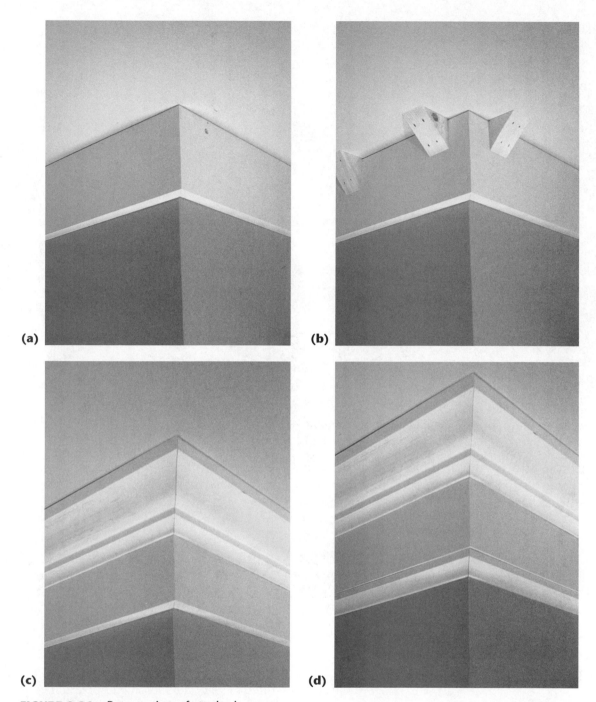

(a)

(b)

(c)

(d)

FIGURE 8-31   Progression of stacked crown

DENTIL CROWN MOLDING

The methods discussed so far can be applied to the installation of dentil crown molding. The main concern when dealing with a molding such as dentil crown, or any molding that has a design or pattern, is to match the design where two pieces meet, particularly at outside corners. (Sometimes you have to let inside corners just fall where they may.) The second piece needs to match the first piece so that they mirror one another. Choose how you want the outside corners to look, and then make sure all the outside corners look the same. **Figure 8-32** shows two choices when it comes to outside corners. Both styles have been used throughout the history of architecture.

In some cases, crown molding may have to be spliced in order to achieve outside corners that

FIGURE 8-32   Two different options for installing dentil crown at an outside corner

look the same. Dentil crown molding has the most pronounced design of any trim that has a repeating pattern. Patterns on embossed trims, such as egg and dart, are not as pronounced or as noticeable; however, their patterns should be kept uniform as well.

SUMMARY

- There are many different types of crown molding, as well as different milling patterns, profiles, and widths.
- The methods for installing all types of crown molding are basically the same.
- Crown installation works best if you work from left to right around a room.
- To measure longer runs of crown, start in the center of a wall, measure from each end, and then add the two measurements.
- Crown does not sit at a true 45-degree angle.
- For the widely used method discussed in this chapter, crown is always cut upside down on the miter saw.
- The finish carpenter can lessen the difficulty of remembering the direction of needed angles by reversing symbols when recording the measurements for crown.
- Use a sample piece of the molding to determine where the bottom of the crown will rest, the proper height mark for the saw, and the size needed for support blocks.
- Gauge blocks or blue chalk lines can be used for marking crown locations on the walls.
- Splices should occur over a ceiling joist or support block.
- Support blocks are usually a must on walls that run parallel to ceiling joists.
- Corners of dentil crown, especially outside corners, need to mirror one another in appearance.

KEY TERMS

compound cuts Cuts made at two different angles.

cornice Both interior and exterior locations where walls and ceilings meet.

crown stops Stops that are mounted onto the miter saw to aid in the cutting of crown molding; they hold the crown in the correct position while the cut is being made.

deadwood Framing that provides something solid to nail to; also referred to as *nailer*; installed during the framing stages of construction.

gauge block A block used when there are many marks of the same length to be made; saves time in not having to repeatedly use the tape measure; can be cut from scrap material.

joists Horizontal framing members that make up the ceiling; wallboard is nailed to the ceiling joists.

milling Sending a piece of trim through a machine to be planed, cut, or given a profile.

nailers Also referred to as *deadwood*; framing whose sole purpose is to provide something solid for wallboard or trim to be nailed to.

support blocks Triangular-shaped blocks that give crown something solid to nail to; installed at corners and where there are no ceiling joists.

snap lines Snap lines are straight reference lines created by using a chalk box. The line serves as a reference by which to cut material or as a line on which material can be installed; trim carpenters use *blue* or *white* chalk only because red chalk bleeds through paint.

undercutting When material is removed from the backside of a piece as it is being coped; allows the crown to seat properly with the adjoining piece of trim.

REVIEW QUESTIONS

1. Is a 2¾-inch crown measured for or cut differently than a 4¼-inch crown?

2. In which direction should one go when working with crown molding?

3. What would you do if you had no ceiling joist to which to nail the crown?

4. How do you determine the size needed for crown support blocks?

5. Explain how you would determine where the bottom of the crown is going to rest on the wall.

6. How would you measure for a long run of crown?

7. If the crown has to be spliced, where should the splice occur?

8. Does crown sit at a true 45-degree angle?

9. Who determines whether or not support blocks are needed?

10. What type of crown is characterized by having evenly spaced blocks that are located at or near the bottom portion of the molding?

11. What is it called when one of the pieces that make up an inside corner is cut at a 45-degree angle, after which its profile is cut out so that the two inside pieces will meet properly?

12. What method is used to join crown that meets at an inside 90-degree corner?

13. How can tracing the profile of trim be helpful when coping?

14. When installing crown, leaving the ends of two joining pieces loose will enable you to do what?

15. If a section of crown is not long enough to go from wall to wall, what should you do?

16. What is meant by the term *stacked crown molding*?

17. Explain the methods used to end a run of crown.

Ceiling Beams

OBJECTIVES

After studying this chapter, you should be able to

- Do framework necessary for ceiling beams
- Determine angles and ceiling slope
- Convert ceiling slope to degrees
- Cut a pattern for ceiling beams
- Assemble and install ceiling beams
- Install tongue and groove

INTRODUCTION

Beams can be installed across flat ceilings or run up the **rake** of vaulted ceilings, where they either do or do not tie into a bigger beam. They can encompass fixtures and give another dimension to an otherwise plain ceiling. They can be incorporated with something such as tongue and groove 1 × 6 and wrapped with crown molding or ¼-inch round moldings. Beams can be added to any ceiling, but hopefully they have been planned for well in advance so that the necessary framework is already in place. Ideally, the necessary framing is in place before the sheetrock is put on. ■

TYPES OF CEILING BEAMS

Structural ceiling beams are set into place as part of a building's structure. They are engineered, designed, and built in accordance with building codes and regulations. They are load-bearing, which means that they hold weight. Certain building designs would not be possible if not for structural ceiling beams.

Nonstructural ceiling beams are not intended or designed for any type of load-bearing purposes (**Figure 9-1**). Their presence is strictly to enhance a ceiling's appearance. To most people, these *false beams* appear just as important to the building's design as the structural beams. Usually there is no notable difference between structural and nonstructural ceiling beams, since they are normally finished with similar materials, which makes everything flow together.

MATERIALS

For ceiling beams that are going to be painted, MDF can be used for both the sides and the cap. If beams are going to be stained, a stain-grade plywood can be used for the sides. A solid wood of the same species used throughout the rest of the home should be used for the cap so that there are no edges exposing layers of veneer. Having a cap of solid wood also means that the edges can be shaped with a router (if called for).

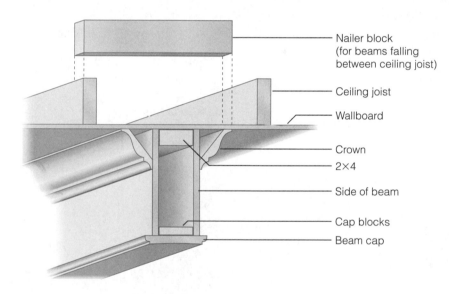

Nailer block (for beams falling between ceiling joist)

Ceiling joist

Wallboard

Crown

2×4

Side of beam

Cap blocks

Beam cap

FIGURE 9-1 Nonstructural ceiling beams

PROCEDURES: CONSTRUCT CEILING BEAMS

 SAFETY TIPS

- Use safety railings on scaffolds, and never lean out over them.
- Do not roll scaffolding over cords or hoses. Doing so can cause scaffolding to tip over.
- Use only GFCI receptacles.
- Use proper gauge extension cords.
- Make sure all tools and cords are grounded.

The following steps are involved in constructing ceiling beams:

1. Lay out the ceiling by measuring and snapping lines.
2. Attach a 2 × 4 (or 2 × 6) to layout lines (giving a solid base to which the beam can be attached).
3. Attach the sides of the beam to the 2 × 4.
4. Attach the cap.
5. Attach the trim (crown, ¼-inch round, cove, etc.).

Beams require a great deal of preliminary planning in order to get a result that looks professional and well-thought-out. Without proper planning, the results may be relatively disastrous and costly. It is better to take a little more time and do the task correctly the first time than to rush blindly into it and end up having to rip the beams apart and redo them.

9-1 FRAMING FOR CEILING BEAMS

Since beams are usually spaced equally apart, sometimes a beam will fall between two ceiling joists. This is where the extra framing is needed, since wallboard alone cannot support any substantial amount of weight. **Figure 9-2** shows framing that is hidden behind the wallboard. The framing needs to be spaced apart no more than 24 inches on center. If the necessary framing does not exist and the wallboard prevents working from below, the framing

may have to be set in place from above, which means going up into the attic if that is possible.

9-2 CALCULATING LAYOUT AND SPACING

For ceiling beams that are intended to be spaced an equal distance apart, several factors must be known so that an accurate calculation of spacing can be made. Without knowing these things, it is impossible to construct evenly spaced beams. Following are the questions you will need answered:

- What is the length of the ceiling (perpendicular to the beams)?
- How many beams are there going to be?
- How many spaces?
- What will the finished width of the beams be?

Figure 9-3 is an example of a simple way to calculate beam spacing for a room. First, take the length of

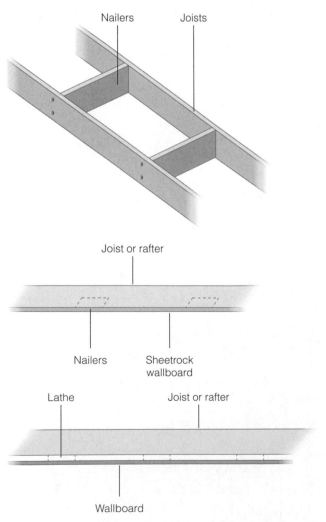

FIGURE 9-2   Framing that supports ceiling beams

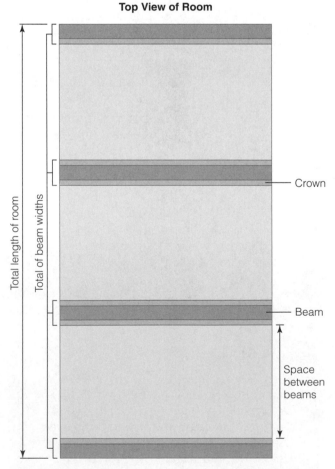

FIGURE 9-3   Diagram of ceiling, with dotted lines to represent ceiling beams

the room and subtract the combined widths of the beams, including allowances for any crown molding that might be used. That number is then divided by the number of spaces, which gives the distance between each beam. To calculate the spacing between beams:

1. Add the combined width of the beams (including crown).
2. Subtract the total beam widths from the length of the room.
3. Divide by the number of spaces.

Put into equation form, this process is expressed as

(length of room) − (total of beam widths)
    ÷ (number of spaces) = space in between beams

Once the correct spacing has been determined, the ceiling can be laid out and marked so that lines can be snapped. Once the ceiling has been laid out, double-check the spacing to make sure everything has been done correctly. The same calculations will work for rooms with any number of beams and for rooms that have intersecting beams, such as the one shown in **Figure 9-4**.

Looking at the profile of a typical beam (refer to Figure 9-1), notice how everything is assembled around the 2 × 4. Chalk lines need to be snapped to represent the 2 × 4. Once all the layout lines have been snapped, the 2 × 4s can be nailed or screwed in place. Make sure that the nails or screws hit the joists or nailers. Once the 2 × 4s have been nailed into place, the sides can be nailed onto the 2 × 4s. If a splice is needed, use a scarf joint (**Figure 9-5**).

Most of the time, an exception exists to every rule. This situation is no different. An exception to

the rule of equal spacing for ceiling beams is shown in **Figure 9-6**. The beams encompass the lights and medallion—fixtures that were preexisting. The beams were an afterthought that came about long after construction had been completed. The main focal point in this case is the combination of the lights, medallion, and beams. These three things had to go together without any eye-catching flaws in the spacing and had to be square in relation to one another. The beams around the lights were constructed without consulting any other part of the ceiling. The unit may or may not be running squarely with the rest of the ceiling, the borders of which are much less important in this case.

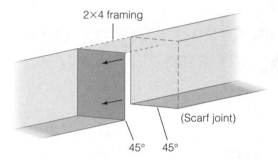

FIGURE 9-5   Scarf joint on ceiling beam

FIGURE 9-6   Beams encompassing fixtures

FIGURE 9-4   Intersecting ceiling beams

> **NOTE**
> If a choice must be made between having the beams square in relation to fixtures or having them square with an existing ceiling that is out of square (as in Figure 9-6), common sense dictates choosing the less common alternative. In Figure 9-6, having the beams out of square with the lights and medallion would have been far more noticeable than having them out of square with the borders of the ceiling. ■

9-3 INSTALLING BEAMS ON VAULTED CEILINGS

When you are installing beams on a vaulted ceiling (**Figure 9-7**), the angle or slope of the ceiling must first be determined so that materials can be cut with the proper angles. The easiest way of determining the ceiling slope is to do the following:

■ Make a level (horizontal) line that meets with the angle of the ceiling.

■ From where the ceiling and level line meet, measure over one foot and make a plumb line all the way up to the ceiling.

■ Measure the plumb line to find out the ceiling slope (**Figure 9-8**).

If the measurement is 6 inches, the slope of the ceiling is 6-12 (which means that for every 12 inches of run, there will be 6 inches of rise). If the measurement had been 7 inches, the ceiling slope would have been 7-12, and so on.

Next, use a speed square to find out what the slope is equal to in degrees. Knowing the degree is necessary, especially since the miter saw uses a scale of degrees, not slope. To determine the slope in degrees, lay the square on any straight edge (a board works fine). Make sure the pivot point of the square is against the straight edge, and then rotate the square until the 6 on the common gauge is in line with the straight edge. A degree mark on the square should now be in line with the straight edge as well (**Figure 9-9**). A 6-12 slope is 26½ degrees. Finding the degrees for any slope is just that simple.

FIGURE 9-7 Beams on a vaulted ceiling

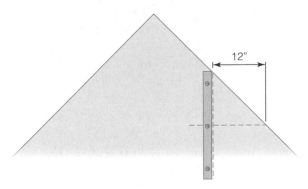

Determining slope or pitch of a ceiling.

(a)

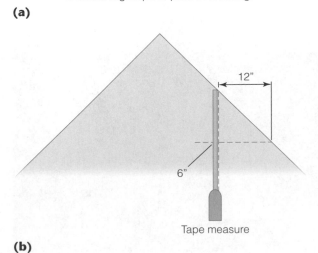

Tape measure

(b)

FIGURE 9-8 Determining ceiling pitch

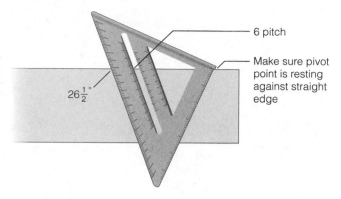

FIGURE 9-9 Finding the degree of ceiling pitch using a speed square

6 pitch

Make sure pivot point is resting against straight edge

$26\frac{1}{2}°$

9-4 CUTTING A PATTERN

Sometimes in carpentry it is helpful to use a pattern. Making ceiling beams is one of those times. A pattern will help you

■ Check certain areas to ensure that the angle is correct and will work properly (**Figure 9-10**).

■ Obtain a correct measurement in places that are often difficult to measure, such as in the tight corner of an angle (**Figure 9-11**). Since the tape measure will fit only so far into the obscure corner, a pattern will ensure accuracy in a situation like this.

■ Mark the correct angle when there are many boards to mark and cut, saving you the time it would take to repeatedly use the speed square (**Figure 9-12**).

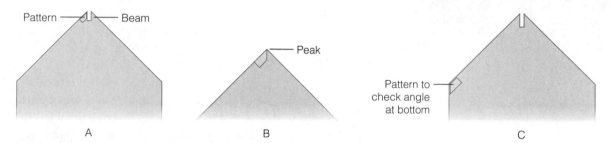

Pattern — Beam

Peak

Pattern to check angle at bottom

A   B   C

FIGURE 9-10 Using a pattern to see if angled cut will fit properly

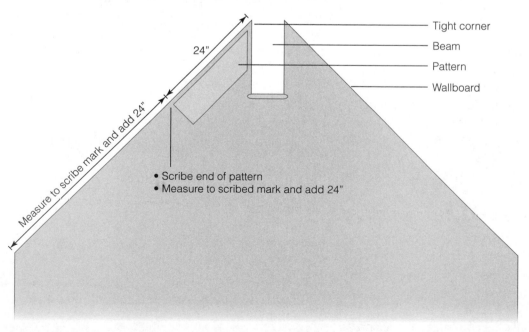

Tight corner

Beam

Pattern

Wallboard

24"

Measure to scribe mark and add 24"

• Scribe end of pattern
• Measure to scribed mark and add 24"

FIGURE 9-11 Patterns assist in obtaining difficult measurements, such as in corners with tight angles.

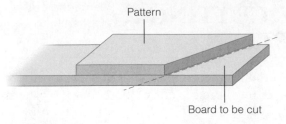

FIGURE 9-12 Using the pattern to mark other boards

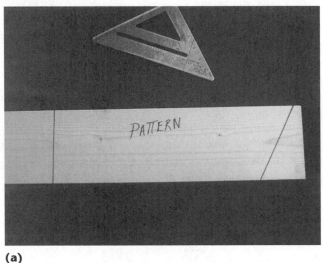

(a)

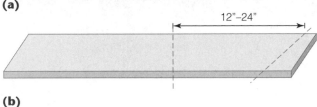

(b)

FIGURE 9-13 Making a pattern that is 12 to 24 inches long

Figure 9-13 shows how an acceptable pattern is made by cutting the slope or degree at one end of the board. The length of the pattern should be between 12 and 24 inches. A pattern that is too short can be inaccurate, and a pattern that is too long is not very practical, as it is heavy and awkward to work with.

9-5 PREPARING AND ATTACHING SIDES AND CAP

The sides of the beam are ripped on the table saw at the desired dimension. The edges are usually left square. After the sides of the beam have been ripped on the table saw and installed, a cap can be made and attached to create the bottom of the beam.

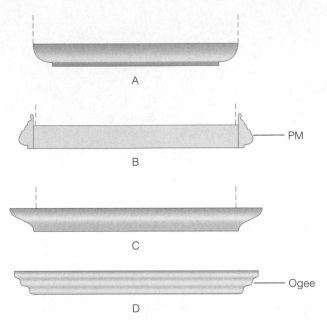

FIGURE 9-14 Cap profiles

Figure 9-14 shows a few cap profiles and situations. A pattern can be used to obtain a measurement for the cap. Nailer blocks can be nailed onto the cap. The blocks will help create a solid and tight fit and will keep the sides in their proper place. The nailer blocks need to be the same width as the 2 × 4. Space the blocks at every 16 inches on center. Cut a gauge block to properly center the cap blocks (**Figure 9-15**). If the cap is spliced at any point, a block is necessary over the splice (**Figure 9-16**). Once the blocks have been nailed on, the cap can be fitted beneath the sides (**Figure 9-17**). Before nailing, make sure the fit is tight and free of gaps.

This method can be used for creating any type of ceiling beam that is desired. The basic procedure is as follows:

1. Rip the desired width for the cap on the table saw.
2. Sand or plane off saw marks.
3. Rout the profile onto the sides of the cap (if called for).
4. Attach the blocks to the cap.
5. Install with finish nails or trim screws.

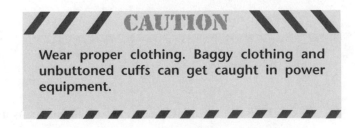

CAUTION

Wear proper clothing. Baggy clothing and unbuttoned cuffs can get caught in power equipment.

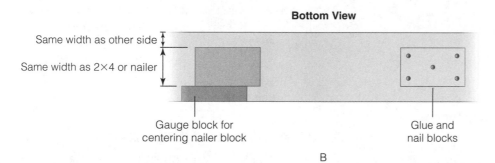

Blocks

16"

Cap (Side View)

A

FIGURE 9-15 Centering cap blocks using a gauge block

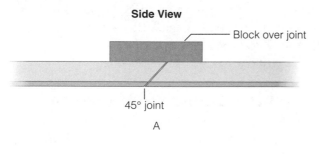

Bottom View

Same width as other side

Same width as 2×4 or nailer

Gauge block for centering nailer block

Glue and nail blocks

B

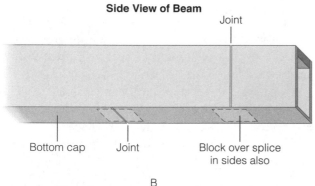

Side View

Block over joint

45° joint

A

Side View of Beam

Joint

Bottom cap   Joint   Block over splice in sides also

B

FIGURE 9-16 Use a block to reinforce splices in cap.

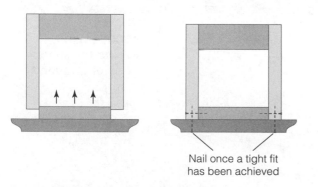

Nail once a tight fit has been achieved

FIGURE 9-17 Fitting the cap beneath the sides, with cap blocks already in place

9-6 INSTALLING TONGUE AND GROOVE

Tongue and groove (T&G) is sometimes incorporated with ceiling beams. Some carpenters prefer to install T&G material first so that the beams can be attached to something solid. Tongue and groove has a tongue side and a grooved side so that the pieces can lock together and support one another (**Figure 9-18**). The T&G is nailed through the tongue so that the nail is hidden; this method is called **blind-nailing**. Starter courses of T&G sometimes have to be **face-nailed**, which means that the nail head is exposed.

Starting Course of Tongue and Groove

The **starter course** of T&G will dictate how the rest of the boards are run. Doing the following will ensure a straight starter course:

- Measure and snap a starter line (**Figure 9-19**).

- Rip the starter pieces on the table saw if installing beams on a vaulted ceiling (**Figure 9-20**).

- Use the starter line as a guide when nailing on the starter course.

The groove side of the board needs to be cut at the same angle as the ceiling, preferably at a greater angle to ensure a tight fit against the wall (Figure 9-20). Cutting the starter pieces at least 5 degrees greater than the angle of the ceiling should be enough to make a tight fit. Trim is sometimes installed between the wall and the beginning course of T&G, which further enhances the appearance of the ceiling treatment.

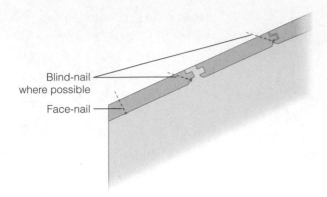

FIGURE 9-18   How tongue and groove joins together

Top View of a Room

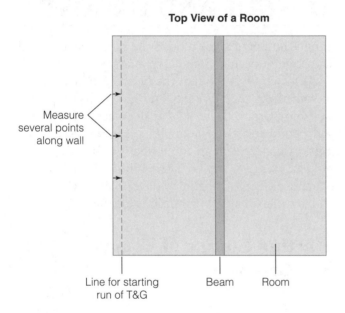

Measure several points along wall

Line for starting run of T&G   Beam   Room

FIGURE 9-19   Snapping a starter line

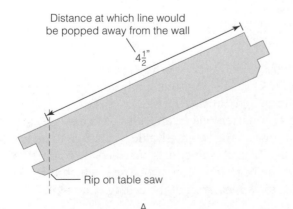

Distance at which line would be popped away from the wall

$4\frac{1}{2}''$

Rip on table saw

A

Some finish carpenters prefer to start running the tongue and groove at the center part of the ceiling, as shown in **Figure 9-21**. This ensures that the starting run and the last run will match one another. If starting in the center of the ceiling, find and mark direct center. Use a **spline** to join the starting pieces (**Figure 9-22**). A spline is a thin strip of wood that joins the grooved edges of T&G ceiling material and T&G flooring. Where the boards join, they will need to be face-nailed. All others, except the last course and the starter course, can be blind-nailed.

Cutting Around Fixtures

To cut out for lights and other fixtures, measure in both directions (**Figure 9-23**), or simply hold the piece up where it goes and scribe a couple of marks. Then, use a **template** to mark the area that needs to be cut out (**Figure 9-24**). Cut the scribed area with a jigsaw (**Figure 9-25**).

Finishing Tongue and Groove

Upon reaching the peak of a vaulted ceiling, it may be necessary to measure each piece again, just to make certain the fit will be tight. In some cases, you may need to cut the last course board so that a fit can be made properly. Cutting off the back of the groove lets the board fit over the tongue properly (**Figure 9-26**). This piece will need to be face-nailed.

Once T&G installation has been completed, lines can be snapped for the 2 × 4s, and the beams can be constructed as discussed earlier. Because the T&G is nailed to joists, it is strong enough to support the non-structural ceiling beams.

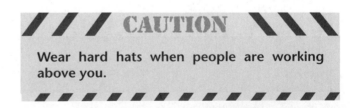

CAUTION

Wear hard hats when people are working above you.

FIGURE 9-20
Ripping the starter piece for a vaulted ceiling

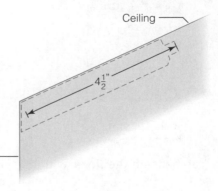

Ceiling

$4\frac{1}{2}''$

Wall

B

Top View of Cieling

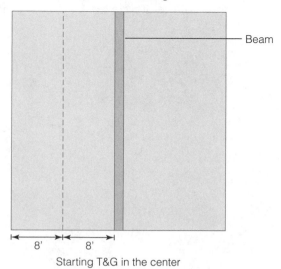

Beam

8' 8'

Starting T&G in the center

FIGURE 9-21 Some people prefer to begin T&G in the center of the ceiling.

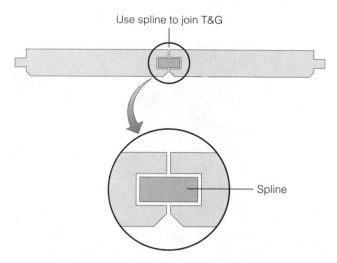

Use spline to join T&G

Spline

FIGURE 9-22 A spline is used to join T&G.

Light

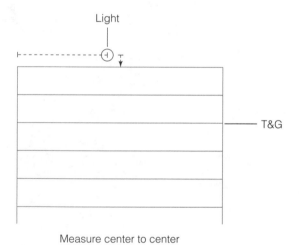

T&G

Measure center to center

FIGURE 9-23 Measuring a light fixture cutout in both directions

FIGURE 9-24 Using a template to mark a fixture cutout

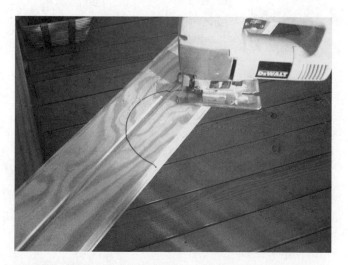

FIGURE 9-25 Cutting out for a fixture using a jigsaw

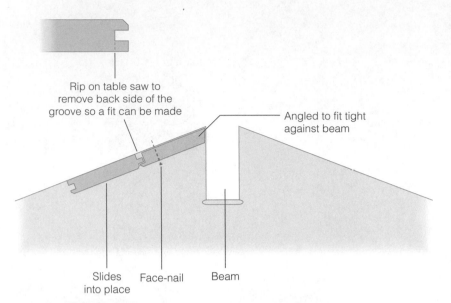

Rip on table saw to remove back side of the groove so a fit can be made

Angled to fit tight against beam

Slides into place   Face-nail   Beam

(a)

FIGURE 9-26 Cutting off the back portion of the groove to fit the final piece

Guard Removed for Visual Purposes

(b)

SUMMARY

- It is best to have the proper framework in place before installing the sheetrock.
- If you will be staining beams, materials used for the beams are usually the same as those used throughout the rest of the home.
- Beams cannot be mounted to sheetrock alone; some sort of framing must be present.
- Framework supporting beams should be spaced at no more than 24 inches on center.
- Beams are usually spaced an equal distance apart.
- Equal spacing of beams can be achieved by making a few simple calculations.

- If installing beams on a vaulted ceiling, you must know the slope in order to make the angles work properly.
- Cutting a pattern is useful when constructing beams on vaulted ceilings.
- Ceiling beams can be incorporated with other ceiling treatments.
- Tongue and groove is nailed with a method referred to as blind-nailing, in which the nails or nail holes cannot be seen.

KEY TERMS

blind-nailing Nailing in the tongue portion of T&G so that the nail is hidden.

face-nailing Nailing to the face side of material.

nonstructural ceiling beams Beams that are not designed for any load-bearing purposes.

rake The slope or pitch of the ceiling.

spline A thin strip of wood that joins the grooved edges of T&G ceiling material and T&G flooring.

starter course The beginning run of material (siding, T&G ceiling treatment, flooring, etc.).

structural ceiling beams Beams that are designed for load-bearing purposes.

template A pattern that is used when there are many of the same type of cuts to be made. Light fixtures sometimes come with templates so that the material can be marked for cutting; afterwards the fixture should fit precisely.

REVIEW QUESTIONS

1. Is sheetrock alone enough to support beams of any kind?

2. What should be the maximum spacing of framing that will be supporting beams?

3. Is it necessary for beams to be spaced equally apart?

4. Does the exact width of the beam need to be known in order to make spacing calculations?

5. What is one benefit of making a pattern?

6. Tongue and groove is nailed in such a way that nails cannot be seen. What is this method of nailing called?

7. When installing beams on a vaulted ceiling, is it necessary to know the angle of the slope?

8. What is the typical number of beams found on ceilings?

9. Draw the profile of a typical beam and label the parts.

10. Beams that are designed to hold up weight are referred to as what type of beam?

11. If a beam is angled at the top where it meets another beam, will the degree of angle be the same as that at the bottom where the beam meets the wall?

12. If it is necessary to splice materials when constructing a beam, what angles should the splice consist of?

13. When you are reversing the direction of tongue and groove (butting the grooved ends together), a thin strip of wood can be glued and inserted into the groove. What is this thin strip of wood called?

14. When blind-nailing is not possible, how are the pieces nailed?

15. Beams set in the pattern of a tic-tac-toe board are said to be what kind of beams?

CHAPTER

10 Stairs

OBJECTIVES

After studying this chapter, you should be able to

- Determine the angle of a staircase
- Cut inside and outside stringers
- Cut and install treads and risers
- Lay out and install balusters
- Lay out and install handrails
- Make a temporary handrail
- Install newels, rosettes, and volutes

INTRODUCTION

Stairs make it possible to get from one level of a structure to another **(Figure 10-1)**. Some terms used to describe stairs are *stairway, flight of stairs, staircase,* and *run of stairs.* Manageable steps consist of a rise and run, which in a series creates a staircase that joins two floors at an incline that can be ascended or descended with ease and safety. Handrails add to the safety and beauty of a staircase. **Balusters** extend vertically from the bottom of the handrail to the top of the stair treads or to the top of a knee wall. The handrail ends by joining to sturdy vertical members known as **newels.** ■

FIGURE 10-1   A finished staircase

Certain variables make each staircase a little different, which makes each one a custom job.

The preferred angle of a staircase is between 30 and 35 degrees (**Figure 10-2**). Any angle greater than this will make the act of climbing stairs strenuous on the body and somewhat dangerous. With the methods covered in this chapter, you should be able to trim any staircase, regardless of its degree of incline.

Building codes (which may vary from state to state) generally restrict a straight run of stairs to a height of 12 feet. A staircase of more than 12 feet must have a platform or a landing at a midpoint. A landing provides a resting point and is a safety feature in the event of a fall. Stairs should be a minimum width of 3 feet, which allows for the passage of two persons traveling in different directions. The ideal width for a staircase is between 3 feet 2 inches and 3 feet 4 inches (**Figure 10-3**).

Figure 10-4 shows some different types of staircases (open, closed, straight run, long L, wide L, wide U, double L, and narrow U). Different types of staircases make it possible to fit a particular staircase to a building's design. Structures of limited space benefit from U-type staircases and winding stairs (both closed and open string).

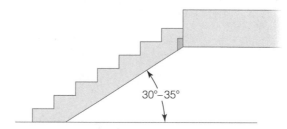

FIGURE 10-2   The preferred angle of a staircase is between 30 and 35 degrees.

TYPES OF STAIRS

Stairs leading down into a basement are called *service stairs*. Stairs joining two finished floors are called *main stairways* or *finished stairways*. Handrails, decorative balusters, and newels, which are typically manufactured, are cut and joined on site by a finish carpenter. Stair parts have to be custom-fitted and must join properly, which requires much skill and experience.

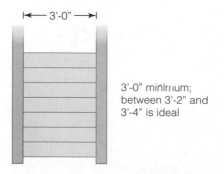

3'-0" minimum; between 3'-2" and 3'-4" is ideal

FIGURE 10-3   The ideal width for a staircase is between 3 feet 2 inches and 3 feet 4 inches.

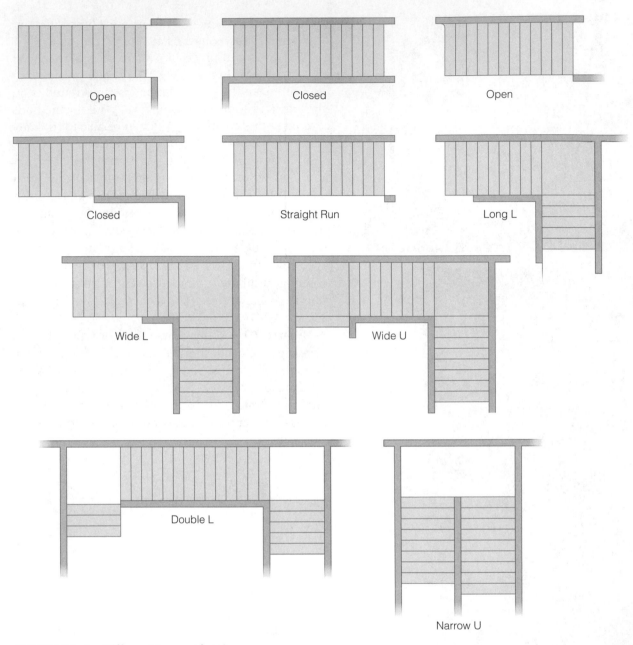

FIGURE 10-4 Different types of stairs

MATERIALS

If the staircase is going to receive a stain finish, all components should be of the same species of stain-quality wood used in the rest of the house. Treads and risers and handrails are typically made of solid oak, both for the sake of appearance and for the purpose of strength.

Balusters are sometimes painted, while handrails and treads and risers are stained. If balusters are to be painted, they can be of a paint-grade material such as poplar or birch. If they are to be stained, the balusters need to be of the same wood species as the handrails and treads.

PROCEDURES: FINISH (TYPICAL) STAIRS

 SAFETY TIPS

- Keep the workspace clean. Debris left on stairs can cause accidents.
- Wear hard hats when people are working above you.
- Pay strict attention when operating tools.
- Hold tools firmly and in a safe manner.

10-1 INSTALLING STAIR STRINGERS

Calculating Stair Stringers

Stringers are cut using 2-inch by 12-inch material and are installed during the *framing* stages of construction. The stringers support the treads and risers. Treads and risers are attached to the stringers. **Treads** lay horizontal and are the part of the staircase that actually gets walked on as a person ascends or descends the staircase. **Risers** are vertical members that connect with the treads. **Unit rise** describes the height of one riser. **Unit run** describes the width of one tread. **Total rise** is the total of all the risers. **Total run** is the total width of all the treads (**Figure 10-5**). It is important to know how the staircase is calculated and constructed in order to fully understand what is, in most cases, a major focal point of a home or building.

Assuming that there are two floors with nothing connecting the two, we will look at how to calculate risers of equal height and come up with a perfect stringer. A perfect stringer is one that has equal risers from the first step to the last. A staircase that has un-

equal risers can cause a fall, so it is important that all of the risers be equal. To calculate the size and number of treads and risers, the total rise is divided by 7:

$$109" \div 7 = 15.57142857$$

Since the number of risers must be a whole number, the number has to be rounded to the nearest whole number. Then that number is divided into the total rise:

$$109" \div 16 = 6.8125 \text{ (which is roughly } 6^{13}\!/_{16}")$$

If you wanted fewer risers, you could divide 109 by 15, which would give you 7.26 (roughly 7¼ inches).

One of two widely accepted rules states that the sum of two risers and one tread should equal 25 inches. The other rule says that the sum of one riser and one tread should equal 17 to 18 inches. Taking into account these two rules, for a riser of 7¼ inches, the tread could be between 10 and 11 inches. If we were going with our first calculation, we could easily use a standard 2-inch by 12-inch for the rough tread (which is actually 11¼ inches wide).

To figure the total run, just multiply the tread width by the number of treads. If there are 10 treads, there will be 11 risers. There will always be one more riser than there are treads.

Once the correct rise has been calculated, the framing square is used to lay out the stringer (**Figure 10-6**). Adjustable setscrews can be purchased for framing squares; they are adjustable and can be tightened at any point along the framing square, making it easier to lay out stringers while yielding a consistent rise and run. Once the entire stringer has been laid out, a circular saw is used to make the cuts. Some carpenters finish the cuts off with a jigsaw; this does away with overcuts, which can cause a stringer to split once weight is applied to it.

A 2 × 4 is sometimes nailed to the stringers. This increases strength and helps eliminate any bouncing movements. For added support at the top of the stringer, cut out for a 2 × 4 ledger strip as shown in **Figure 10-7**.

In Figure 10-7, notice how the bottom of the stringer has been cut 1½ inches shorter than the other risers. Depending on the type of flooring being used, cut the bottom of the stringer in a way that will give you the same amount of rise as the other risers

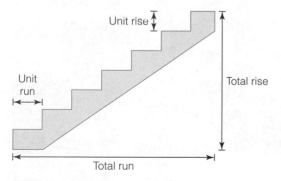

FIGURE 10-5 Rise and run

(a)

(b)

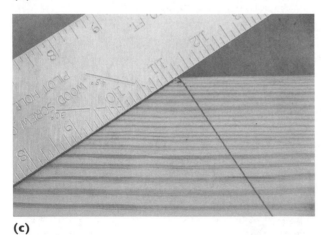

(c)

FIGURE 10-6 The framing square is used to lay out the rough stringer.

(**Figure 10-8**). Once this first stringer has been cut, it can be used as a pattern for cutting the other stringers. Laying out the stringers individually can result in stringers of slightly different dimensions. Install the stringers with 3-inch screws or framing nails.

The calculation and layout principles are the same for all the different styles of stringers. (See **Figure 10-9**, which shows different types of stringers, and the section later in this chapter on types of finished stringers.)

CAUTION

Keep blades on equipment sharp. A dull blade overworks the motor and can be unsafe. Hold the saw with a firm grip while cutting stair stringers. Saws will often kick back when material binds on the blade.

(a)

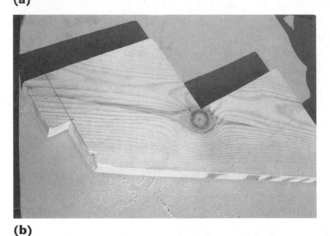

(b)

Stringer resting on 2×4 ledger

(c)

FIGURE 10-7 Notching stringer for ledger strip

FIGURE 10-8 Cutting stringer to allow for height of flooring

10-2 DETERMINING ANGLE OF STAIRCASE

Knowing the angle of a staircase before attempting the finish work is important because a lot of the materials will have to be cut with an exact angle. Handrails, balusters, and even stringers will have to be cut so that their angle matches the angle of the existing staircase. Balusters are decorative members that support the handrail.

Determining the angle of the existing staircase is done by making a plumb line, and then using the speed square to obtain the angle (**Figure 10-10**). It

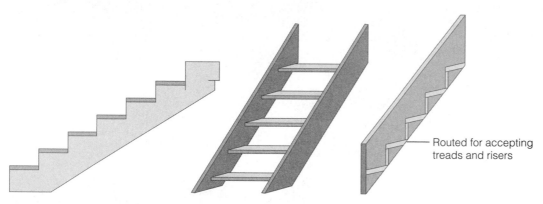

FIGURE 10-9 Different styles of stringers

Routed for accepting treads and risers

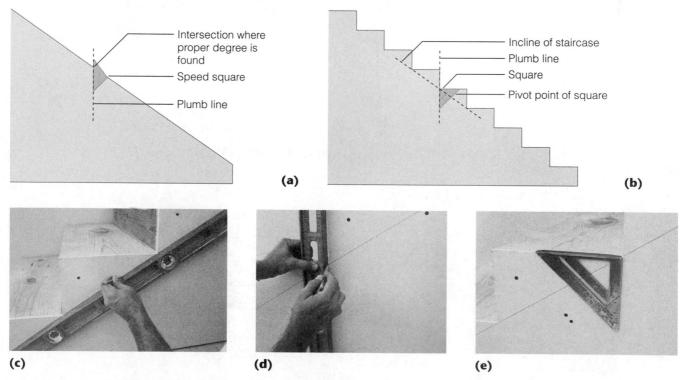

Intersection where proper degree is found
Speed square
Plumb line

(a)

Incline of staircase
Plumb line
Square
Pivot point of square

(b)

(c)   **(d)**   **(e)**

FIGURE 10-10 Determining the angle of the existing staircase

will be necessary to refer back to this angle quite often during the staircase construction.

Types of Finished Stringers

Unlike the rough stringer (or carriage), the finished stringer is usually ¾-inch thick, and it will be part of the exposed trim work on the staircase. If the stair system is going to be stained, the finish stringer should consist of a stain-quality material, as it will be just as visible as the balusters and handrails.

Following is a list of typical finished stringers:

- Open stringer (**Figure 10-11**)
- Mitered stringer (**Figure 10-12**)
- Closed stringer (**Figure 10-13**)
- Housed stringer (**Figure 10-14**)
- Straight stringer (**Figure 10-15**)

Theoretically, finished stringers can be laid out with the same method used for laying out rough stair stringers (by using a framing square); however, to ensure a good custom fit of the finish stringer, other methods can be used.

Open stringer. The open stringer can be set into position against the staircase, and the outline of the

treads and risers can be scribed lightly onto the stringer. Afterward, with a framing square positioned on the scribed lines, you can scribe a darker and more definite line so that the cut can be made true and square. Retracing with the framing square solves the problem of the occasional riser being out of plumb.

Another method used for marking the stringer is to use a **preacher**. The preacher fits over the stringer

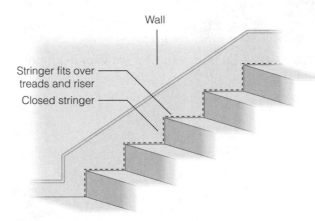

FIGURE 10-13   Closed stringer

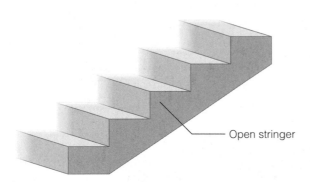

FIGURE 10-11   Open stringer

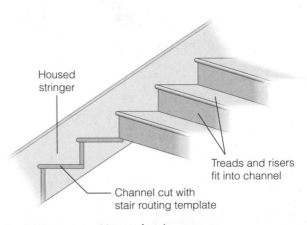

FIGURE 10-14   Housed stringer

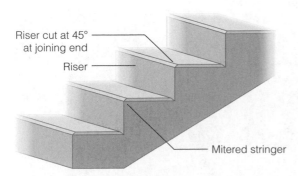

FIGURE 10-12   Mitered stringer

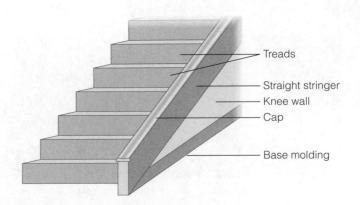

FIGURE 10-15   Straight stringer

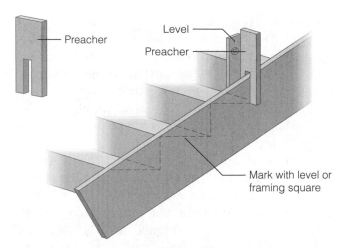

FIGURE 10-16 Marking treads with a preacher and a framing square

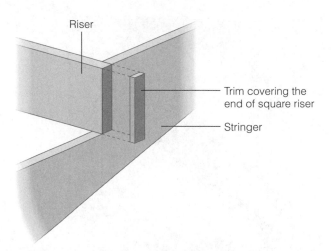

FIGURE 10-17 Trim covering the exposed end of a square-cut riser

and sits on the tread portion of the carriage. Once the preacher is plumbed with a level, a line can be scribed onto the stringer. Mark the bottom portion of the preacher for an accurate tread mark. Once again, using a framing square will yield the correct tread mark, since the line made with the preacher is plumb (**Figure 10-16**).

> ### NOTE
>
> With either method, the stringer must be held securely in place throughout the process of marking each tread and riser. Any movements will result in an inaccurate stringer. ■

If you are cutting the stringer with square cuts, it is important to realize that (after the treads and risers have been installed), another piece of trim will have to be added to cover the end of the riser (**Figure 10-17**). If a mitered look is preferred, you will have to create a mitered stringer.

Mitered stringer. The methods for making a mitered stringer are the same as those used for the open stringer, except that the riser cuts will have to be cut at a 45-degree angle to accommodate a riser that also has a 45-degree cut. The mitered stringer can be marked using methods similar to those used for the open stringer. The riser portions of this stringer are cut at a 45-degree angle so that the stringer and the riser are joined tightly, eliminating the need for any trim at that particular point.

Closed stringer. The closed stringer is set on top of the carriage and secured to the wall so that layouts can be made without moving the stringer. Typically risers

are installed first so that the stringer can be laid out to fit over them. Use a level to mark the tread and riser marks onto the stringer (**Figure 10-18**). Afterward, cut the stringer along the layout lines. The bottom of the stringer should rest on the floor and the top should sit flat on the landing.

The landing portion of the stringer needs to be cut so that a similar baseboard can be mated to it (**Figure 10-19**). The same thing needs to be done at

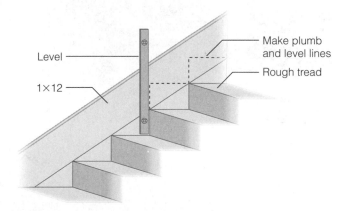

FIGURE 10-18 Using a level to mark treads and risers onto the closed stringer

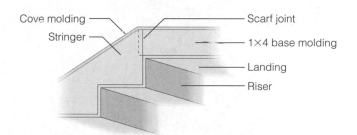

FIGURE 10-19 Landing portion of the stringer meeting base molding

FIGURE 10-20   The top of the stringer meeting base molding

the bottom (**Figure 10-20**). Before installing the stringer, double-check to make sure all cuts fit tight against the risers and to make sure the top (landing) and bottom (floor) sections of the stringer are cut to accurately mate with a similar base treatment. Make any necessary adjustments before installation. The stringer cannot be cut or adjusted once it has been fastened to the wall.

Housed stringer. The housed stringer is made using a stair template. The stair template can be adjusted to different rise and run patterns. A router with a straight fluting bit is used with the stair template to create dadoes into the stringer (**Figure 10-21**). After the dadoes have been cut, treads and risers are cut and fitted into the stringer. Wedges or shims are inserted on the underneath side of the stringer, between the stringer and the tread and riser pieces (**Figure 10-22**). The wedges help tighten and close gaps between the treads and risers and the stringer.

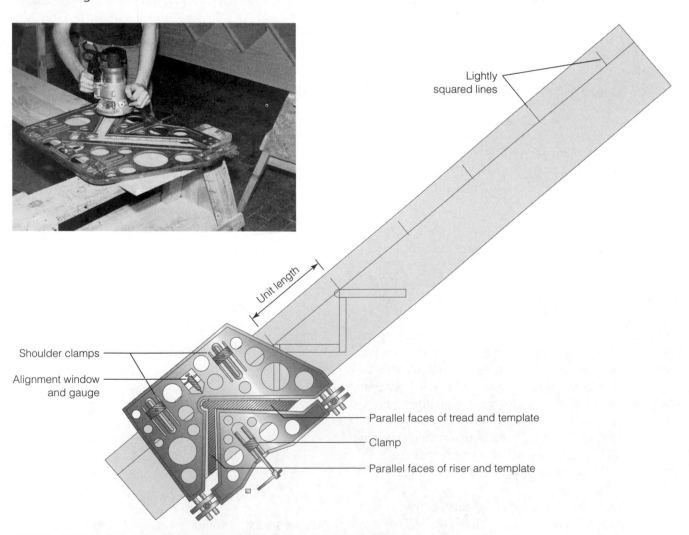

FIGURE 10-21   A stair template is used for cutting a housed stringer.

SAFETY TIP

When using a stair template, follow the manufacturer's instructions and safety warnings.

Straight stringer. The straight stringer is used on stair systems that incorporate a knee wall. Typically the knee wall ascends with the same degree of incline as the staircase. On the stair side of the knee wall, there is a closed stringer, and on the opposite side of

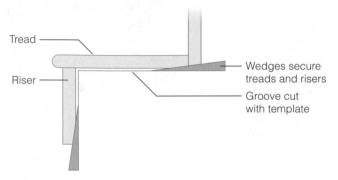

FIGURE 10-22 Wedges are inserted between bottom portions of the treads and risers, securing them tightly against the stringer.

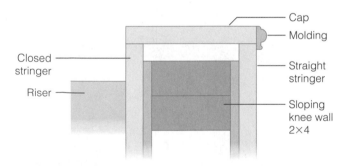

FIGURE 10-23 A cap on top of a knee wall

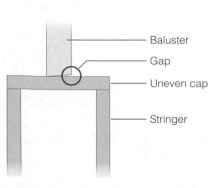

FIGURE 10-24 Balusters will not join properly to a cap that is not sitting flat.

the knee wall there is a straight stringer. The two stringers are joined by a board, sometimes referred to as a cap, which is positioned on top of the two stringers (**Figure 10-23**).

The tops of both stringers should remain at the same height so that the cap is sitting flat. If one stringer is higher than the other, the cap will not sit flat and the balusters will not join to it properly (**Figure 10-24**). The straight stringer is trimmed at the bottom and also where it joins the top board. The bottom and top of the stringer should be cut with the base molding and a top molding in mind (**Figure 10-25**).

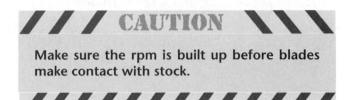

/// CAUTION ///

Make sure the rpm is built up before blades make contact with stock.

10-3 INSTALLING TREADS AND RISERS

Treads and risers are commonly made of oak. Treads need to be made of a solid wood for strength. The tread should have what is called a *nosing*, which overhangs the previous tread. The nosing can have a profile, and sometimes trim is added beneath the nosing (**Figure 10-26**). The amount of overhang for the nosing ranges from 1 inch to 1½ inches off the face of the riser. It is important to note that the nosing is not considered part of the unit run. Treads overhang the finished stringer on open stairs (**Figure 10-27**).

Risers are installed before the treads. Installing the riser first allows for the fastening of the tread through the back of the riser. Glue blocks can be used where the

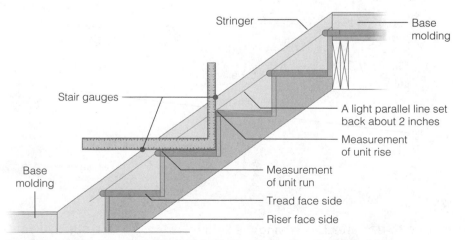

FIGURE 10-25 The bottom and top situations of a straight stringer

riser meets the tread, or a rabbeted joint can be made so that the riser fits into a groove in the tread (**Figure 10-28**). The gap between riser and tread is sometimes concealed with cove molding (Figure 10-26).

When measuring the space for a tread or riser, check the wall with a framing square and make the appropriate adjustments by cutting the tread to fit the space (**Figure 10-29**). Measure each tread and riser individually to ensure a tight fit. If the tread is overhanging the finished stringer (as would be the situation with an open staircase), keep the amount of overhang the same as the nosing so that the tread appears uniform all the way around (**Figure 10-30**).

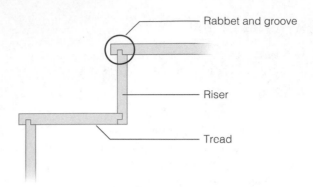

FIGURE 10-28 A rabbetted joint where tread and riser meet

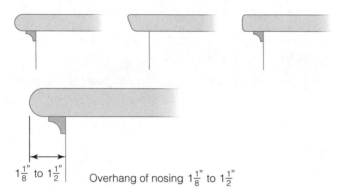

FIGURE 10-26 Tread nosing can have a profile and trim beneath the nosing.

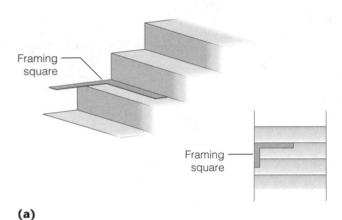

(a)

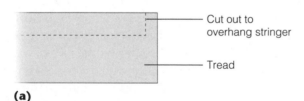

(a)

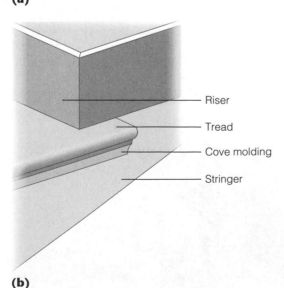

(b)

FIGURE 10-27 Treads overhang the finished stringer on open stairs and are returned back to the stringer.

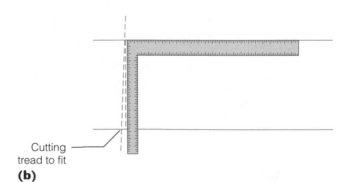

(b)

FIGURE 10-29 Check the wall with a framing square and make corresponding adjustments onto the tread.

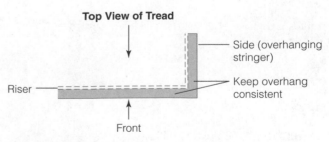

FIGURE 10-30 Keep the amount of overhang consistent.

Cut stair materials with a crosscut blade, and sand any rough edges smooth before installing. If installing treads and risers over rough framing, use construction adhesive to eliminate possible noise due to movement (such as a squeaky stair tread). Fasten treads and risers with finish nails or trim screws. Unless specified otherwise, two fasteners at each rough stringer is sufficient for the treads and risers (**Figure 10-31**).

Occasionally, tread and riser pieces are nailed onto the rough treads and risers, and carpeting is installed between them (**Figure 10-32**). The exposed edges of

the treads and risers are trimmed with molding. The ends of the molding are returned back to the stringer with one of two methods—ending the piece with a 45-degree angle returning to the wall, or ending the piece with a 90-degree angle to the wall (**Figure 10-33**).

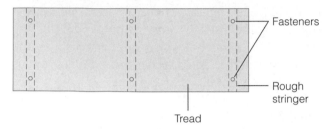

FIGURE 10-31 Fasten the treads and risers to rough stringers.

FIGURE 10-32 Tread and riser pieces with carpeting in between

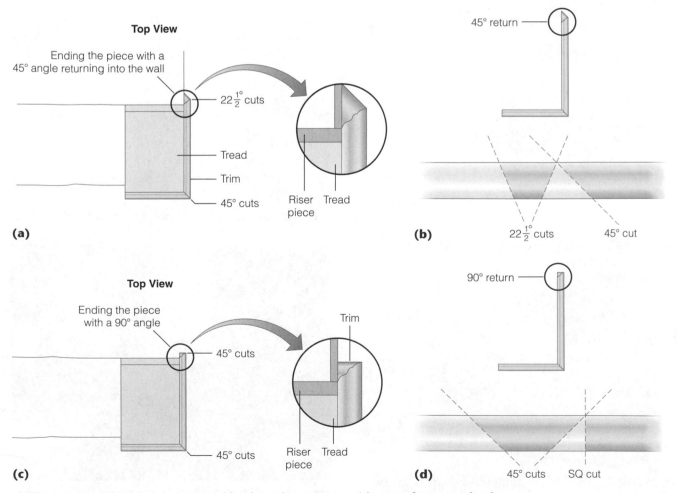

FIGURE 10-33 Molding is returned back to the stringer with one of two methods.

The risers can be mitered to fit the finished stringer (**Figure 10-34**).

After all treads and risers have been installed, treads are laid out, and a hole is drilled for accepting certain types of balusters (**Figure 10-35**). Codes require that the face of the baluster be in line with the face of the riser and that the baluster in between the nosing balusters be centered between the two (**Figure 10-36**).

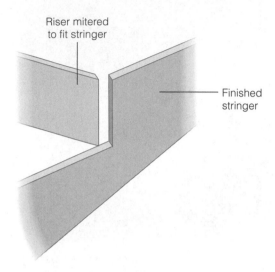

FIGURE 10-34 Risers can be mitered to fit the finished stringer.

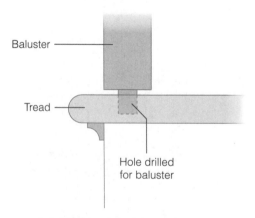

FIGURE 10-35 A hole is drilled into treads for accepting balusters.

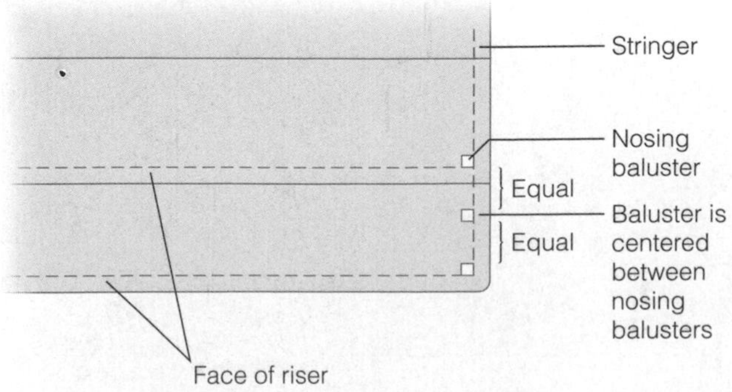

FIGURE 10-36 Baluster is centered between the two nosing balusters.

10-4 INSTALLING THE STARTER STEP

On some staircases, the starting step is rounded on one or both ends, depending on the design (**Figure 10-37**). The step is rounded, for example, when a volute is being used as a beginning point of the handrail. The volute, positioned directly above the rounded portion of the tread, has a radius and is joined to the handrail.

A few types of starter steps are shown in **Figure 10-38**. Starter steps can be ordered from a stair parts manufacturer, or a starter step can be made by transferring the radius of the volute, or a comparable radius, onto the tread stock (**Figure 10-39**). Cut the scribed radius with a jigsaw. Smooth the radius with a sander (palm, belt, or orbital sander), removing all high and low spots in the wood. Use a router with a ¼-inch round bit to round over the edges of the radius and the front edge of the step (**Figure 10-40**).

CAUTION

Stabilize material or stock before cutting. Check safety features on a regular basis to make sure they are in proper working order.

FIGURE 10-37 A rounded starting step

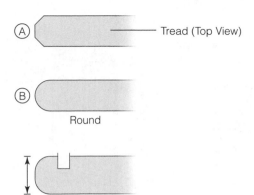

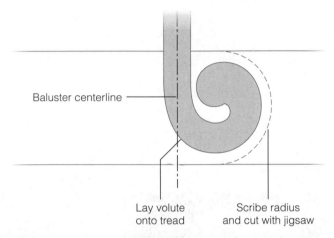

FIGURE 10-38 A few types of starter treads

FIGURE 10-39 Marking radius onto the tread, using volute as a guide

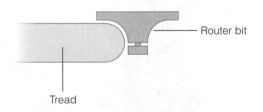

FIGURE 10-40 Using a round bit to rout the starting step

10-5 INSTALLING THE BALUSTRADE

The term **balustrade** refers to the upper members of the staircase. The handrail, newels, and balusters are the members that make up the balustrade. The balusters support the handrail on open staircases. Newels are beginning and ending supports for the handrail. Members of the balustrade are usually factory-made and are cut and assembled at the jobsite. The following sections cover members of the balustrade in detail.

Laying Out the Balustrade

Layouts have to be precise so that the handrail, balusters, and newels join precisely. Following is a list of the steps involved in the balustrade layout:

1. Mark centerline.
2. Mark center of each baluster.
3. Transfer layouts to handrail.

First, mark the centerline, which will indicate the center of all the members (balusters, handrail, newel,

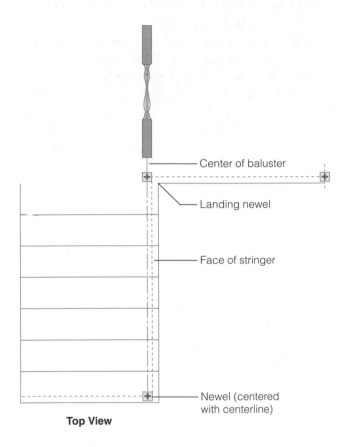

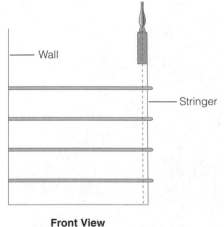

FIGURE 10-41 Layout for baluster

and even rosettes). When marking layout lines, use light pencil marks that can be sanded away easily. On an open staircase, codes state that the edge of the baluster should be in line with the face of the finish stringer, which means that the centerline would be moved over the distance of half the baluster width (**Figure 10-41**). If you are laying out the top of a knee wall or a balcony, the centerline can be run directly in the center (**Figure 10-42**).

Laying Out a Balcony Balustrade

The balcony balustrade is similar to the staircase balustrade in most ways. Balcony newels support the handrail, and balusters are installed vertically beneath the handrail. The handrail height for a balcony is to be no less than 36 inches, as stated by the international residential code. All members of the balcony balustrade are installed with the methods already discussed. Half newels as well as quartered newels and rosettes are sometimes used where balcony handrails end at a wall (**Figure 10-43**).

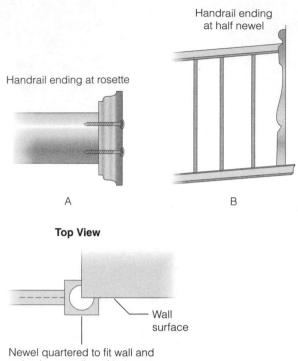

Top View

(a)

(b)

FIGURE 10-43 Balcony handrail ending at quartered newel, rosette, and half newel

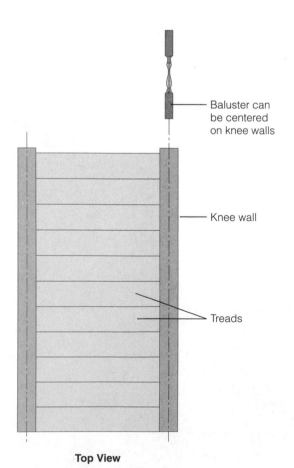

Top View

FIGURE 10-42 The layout on top of a knee wall can be directly in the center.

10-6 INSTALLING HANDRAILS

The handrail is installed for safety reasons and provides a decorative appearance. It runs parallel with the staircase, which means that it ascends at the

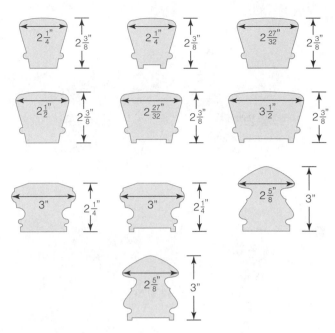

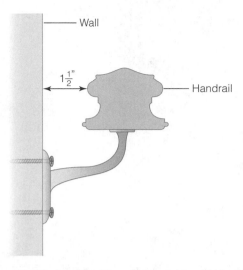

FIGURE 10-46 There should be 1½ inches of clearance between the handrail and the wall.

FIGURE 10-44 Different profiles and dimensions of handrails

same angle as the staircase. Handrails come in many different profiles and dimensions (**Figure 10-44**). Handrails are traditionally made of solid oak, which ensures strength and can be stained. In a closed staircase, the handrail is mounted to the wall with decorative brackets. The brackets are attached to the

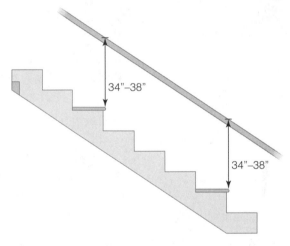

FIGURE 10-47 Handrail height should be between 34 and 38 inches off the nosing of the treads.

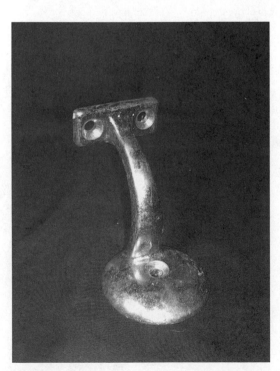

FIGURE 10-45 Brackets for attaching handrail to wall

handrail with screws, and longer screws secure the brackets to the wall studs (**Figure 10-45**).

There should be a continuous 1½-inch clearance between the handrail and the wall for general use and safety purposes (**Figure 10-46**). Codes require that handrails be at a height of 34 to 38 inches from the nosing of the treads (**Figure 10-47**). On balconies, the height of the handrail is to be no less than 36 inches. Volutes, turnouts, and easings are sometimes attached to the handrail as a starting or ending point (**Figure 10-48**). (See the section later in this chapter on installing volutes for more information about volutes, turnouts, and easings.) Following is a summary of the required measurements:

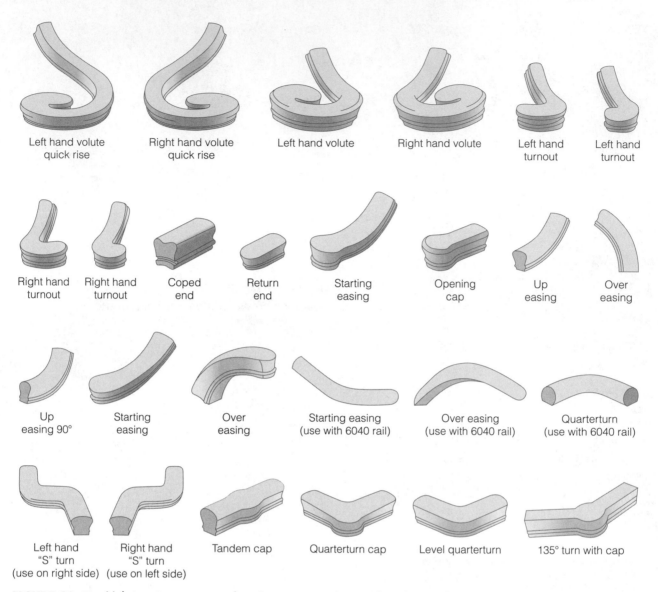

Left hand volute quick rise

Right hand volute quick rise

Left hand volute

Right hand volute

Left hand turnout

Left hand turnout

Right hand turnout

Right hand turnout

Coped end

Return end

Starting easing

Opening cap

Up easing

Over easing

Up easing 90°

Starting easing

Over easing

Starting easing (use with 6040 rail)

Over easing (use with 6040 rail)

Quarterturn (use with 6040 rail)

Left hand "S" turn (use on right side)

Right hand "S" turn (use on left side)

Tandem cap

Quarterturn cap

Level quarterturn

135° turn with cap

FIGURE 10-48 Volutes, turnouts, and easings are starting and ending points for handrails.

- Clearance between wall and handrail: 1½ inches

- Height of handrail: 34 to 38 inches

- Handrail height along balcony: no less than 36 inches

Some handrails are grooved along the bottom so that square top balusters can be fitted into the grooves. After the balusters have been secured in place, **fillets** are cut and inserted in the spaces between the balusters (**Figure 10-49**). The handrail has to be drilled so that it can accept round top balusters (**Figure 10-50**).

The handrail should be laid out carefully when using any type of baluster; otherwise, the members will not be plumb and will not have the appearance of being uniform. The layout of the handrail depends on

the layout of the balusters. (See the sections that follow on laying out the handrail and laying out the individual baluster.) When handrails end at a wall, a rosette is sometimes used (**Figure 10-51**).

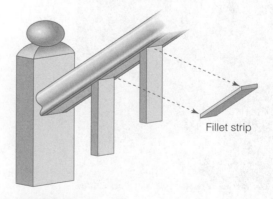

Fillet strip

FIGURE 10-49 Fillets are inserted between balusters.

(a)

FIGURE 10-51   Handrail ending with a rosette

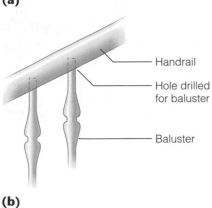

Handrail

Hole drilled
for baluster

Baluster

(b)

FIGURE 10-50   Handrails must be drilled to accept round top balusters.

Laying Out the Handrail

Layout of the handrail can be performed in a number of ways. The handrail can be laid directly on top of the treads and a square can be used to mark each individual baluster location (**Figure 10-52**). The newel location can also be scribed onto the handrail, at which point the handrail would be cut off. Once all baluster marks have been made, use a square to continue lines across the bottom of the handrail. Where those lines intersect with a centerline is where the center of the baluster will be located. (For step-by-step instructions on drilling the handrail for accepting round top balusters, see the next section.)

Another method for laying out the handrail is to make a temporary handrail using a 1-inch by 4-inch board. Making a temporary handrail (TH) will help with a number of tasks. Set the TH at the same incline and the same height intended for the handrail. Use a level to plumb up from the baluster layouts and scribe them onto the TH (**Figure 10-53**). The TH can be used to obtain exact measurements for each individual baluster (**Figure 10-54**). The newel locations can also be scribed onto the TH, since the newel height can be determined at this point

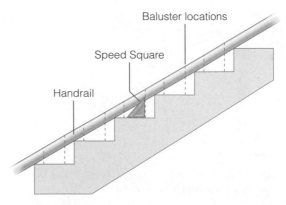

Baluster locations

Speed Square

Handrail

FIGURE 10-52   Laying out handrail, directly on top of the treads, with a square

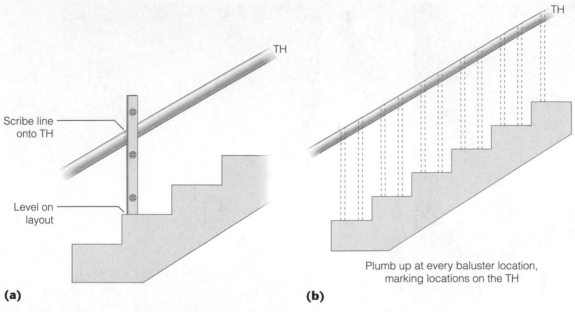

(a)

(b)

Plumb up at every baluster location,
marking locations on the TH

FIGURE 10-53   Use a level to scribe baluster layouts onto a TH.

(**Figure 10-55**). (See also the later section on making a temporary handrail.)

After the TH has been marked, the layouts can be transferred onto the handrail (**Figure 10-56**). Balusters can be cut with the measurements obtained by using the TH. Numbering the balusters will help maintain the correct order during installation.

> **NOTE**
>
> Due to imperfect stringers, each baluster measurement may vary; therefore, individual measurements should be taken. ∎

Drilling Handrail for Accepting Round Top Balusters

Holes must be drilled into the handrail so that the tops of the balusters are recessed into the handrail. The angle of the staircase must be known so that an exact pattern can be cut. The pattern will ensure that holes are drilled properly and will accept the plumb balusters. If the staircase is 35 degrees, cut a pattern with a *long 35* or the *back cut* of a 35-degree angle by subtracting 35 from 90 degrees to find the proper degree.

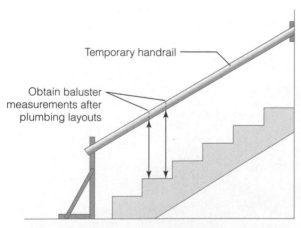

Make sure TH is the same dimension as the actual handrail, and make sure it is set at the proper height

FIGURE 10-54   Obtaining exact baluster measurements using the TH

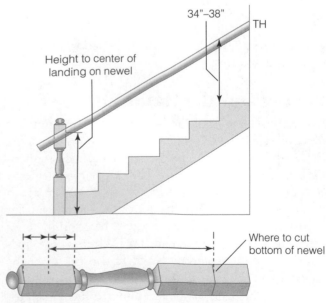

FIGURE 10-55   Figuring the newel height using the TH

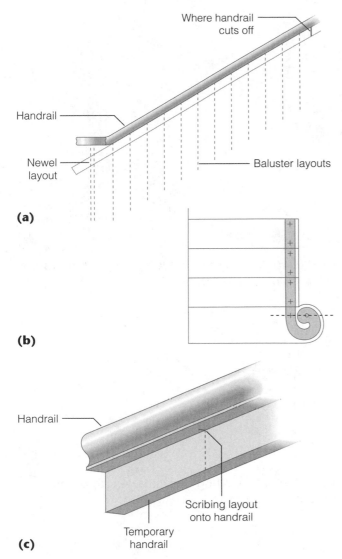

(a)

(b)

(c)

FIGURE 10-56 Transferring layouts from the TH to the handrail

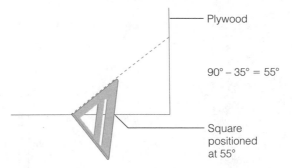

FIGURE 10-57 Cutting a pattern

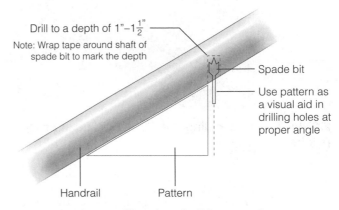

FIGURE 10-58 Drilling handrail for round top balusters, showing depth

Cut the pattern at 55 degrees (**Figure 10-57**). (The same method would be used for staircases with other angles, such as one that is 30 degrees.)

Situate the handrail and pattern on a work surface, and drill at each baluster location. Using a bit the same size as the baluster top, drill about 1 inch to 1½ inches into the handrail (**Figure 10-58**). Assistance may be needed in lining up the balusters with the holes. When all of the baluster tops are seated fully into the handrail, fasten them with finish nails through the side of the handrail at each baluster location. Before nailing, make sure the handrail is straight and does not have a sag (**Figure 10-59**).

Making a Temporary Handrail

Usually a 2-inch by 4-inch bracing is used at the bottom and top of the staircase, but a 1-inch by 4-inch

bracing can be used for the temporary handrail (TH). Rip the TH to the same vertical dimension as the handrail being used so that actual baluster measurements can be taken. Secure the TH to the bracing,

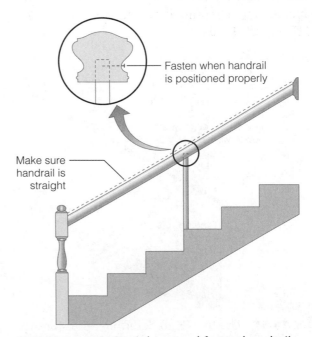

FIGURE 10-59 Straighten and fasten handrail

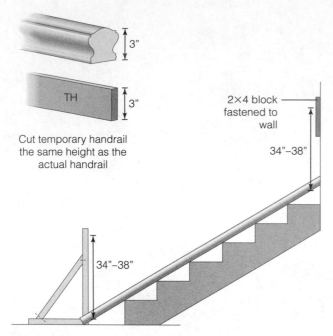

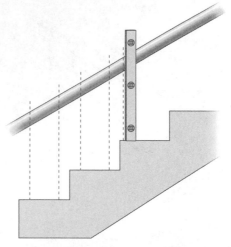

FIGURE 10-62   Use a 4-foot level to plumb up to the TH at each baluster location.

FIGURE 10-60   Securing the TH to bracing at proper height

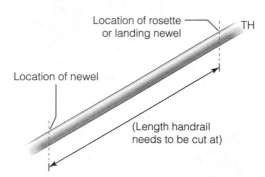

FIGURE 10-63   Getting the actual length of the handrail using the TH

between 34 and 38 inches off the nosing of the treads (**Figure 10-60**). Use the TH to determine the starting newel height (**Figure 10-61**). Use a 4-foot level to plumb up at each baluster location and mark them onto the TH. A center mark can be made in the case of round top balusters (**Figure 10-62**).

Once the baluster layouts have been marked onto the TH, accurate baluster measurements can be taken (**Figure 10-63**). The actual length of the handrail can be obtained by measuring between newel layouts on the TH (**Figure 10-64**).

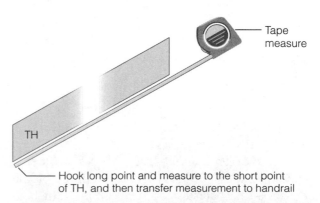

FIGURE 10-64   Measuring between newel layouts on the TH to determine actual length of handrail

10-7 INSTALLING BALUSTERS

Balusters are vertical members that extend from the bottom of the handrail to the stair treads or to the knee wall. The International Residential Code states that a sphere of 4 inches should not be able to pass through

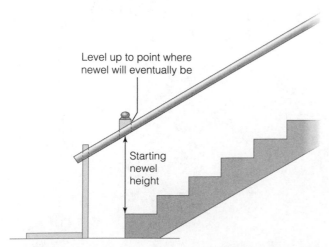

FIGURE 10-61   Use the TH to determine the starting newel height.

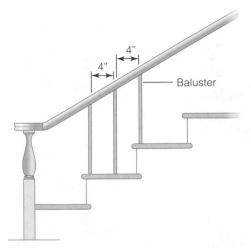

FIGURE 10-65 There should be no more than 4 inches of space at any point between balusters.

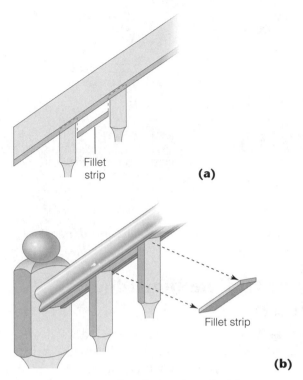

FIGURE 10-67 Fillets being installed between balusters

the balusters at any point (**Figure 10-65**). This spacing is for safety reasons. A spacing of more than 4 inches would allow a small child to pass through and could result in a fall. The code also states that the front edge of the baluster should be in line with the face of the riser, as well as in line with the face of the finished stringer (**Figure 10-66**).

Some balusters have square tops that fit into a milled groove along the bottom of the handrail. The balusters are cut with the angle of the staircase and then glued and nailed into place. Later fillets are installed between the balusters (**Figure 10-67**). Other balusters have round tops that must be fitted into the handrail after the handrail has been laid out and the holes have been drilled for accepting the balusters. The bottoms of most balusters have what appear to be dowel rods extending out ¾ inch. The dowel portion of the baluster is slid into position after the treads have been laid out and holes have been drilled (**Figure 10-68**).

Balusters are attached to a knee wall after the baluster has received an angled cut the same as that of the stair incline. Balusters are sometimes attached directly to the cap of the knee wall; other times they are attached to what is called a **buttress cap** (**Figure 10-69**). A buttress cap is usually attached to a knee wall; it has a milled channel for accepting the bottom of square balusters. After fastening the baluster to the buttress cap, fillets are cut and installed in the spaces between (**Figure 10-70**).

Balusters can be run in several different patterns (**Figure 10-71**). When run one way, the balusters may mirror the incline of the stairs, but when run another way, they may mirror the tread pattern. On some staircases, three balusters per tread are used.

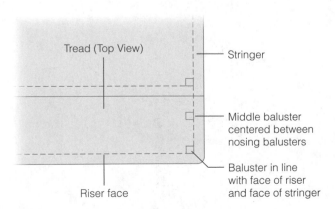

FIGURE 10-66 Baluster placement

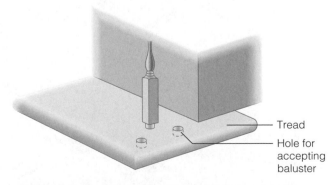

FIGURE 10-68 Dowel portion of baluster in place

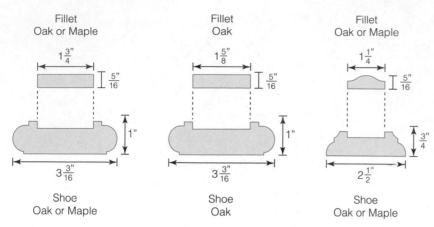

Fillet
Oak or Maple

Fillet
Oak

Fillet
Oak or Maple

Shoe
Oak or Maple

Shoe
Oak

Shoe
Oak or Maple

FIGURE 10-69   Buttress cap profiles and dimensions

Laying Out the Individual Baluster

Next mark the center of each baluster. Codes state that the edge of the baluster should be in line with the face of the riser as well as in line with the face of the finish stringer. The layout for the balusters near the tread nosing would determine where the baluster in between is centered (**Figure 10-72**). If using three balusters per tread, space the two middle balusters evenly between the two nosing balusters. Once all the

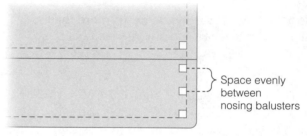

Space evenly between nosing balusters

A. Three balusters per tread

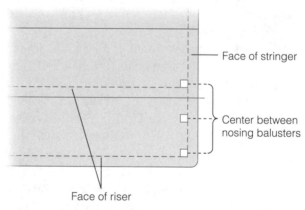

Face of stringer

Center between nosing balusters

Face of riser

B. Two balusters per tread

FIGURE 10-72   A baluster centered between two nosing balusters, with spacing of three balusters per tread

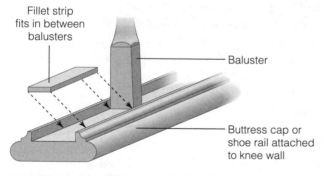

Fillet strip fits in between balusters

Baluster

Buttress cap or shoe rail attached to knee wall

FIGURE 10-70   Fillets installed with buttress cap

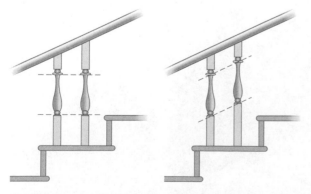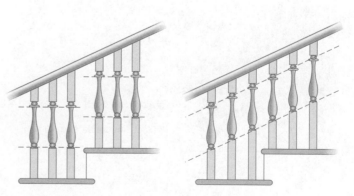

FIGURE 10-71   Patterns in which balusters can be run

baluster centers have been marked, the handrail can be laid out.

10-8 INSTALLING NEWELS, ROSETTES, AND VOLUTES

The following sections explain how to properly attach newels, rosettes, and volutes.

Newels

Newels are sturdy beginning and ending points for the handrail and balustrade. Newels also provide support at intermediate locations such as at landings and along balconies. Newels also serve as transitional points, allowing the handrail to change direction or go from being an ascending member to a horizontal balcony or landing member (**Figure 10-73**). Newels are dimensionally larger than balusters, but they are milled with the same patterns as the baluster so that the balustrade is a matching unit. Newels are mounted to the staircase with either dowels or lag screws, which provide strength and support to the unit. A newel or handrail that has not been mounted properly could result in a fall.

There are beginning newels and balcony newels. Newels with a larger vertical landing (a flat surface on which the handrail is mounted) allow handrails to change in direction and in height (**Figure 10-74**). Some starting newels have a dowel pin in place of the finial, which accommodates a **volute** (**Figure 10-75**). A volute is a decorative piece that is attached to the handrail. Half newels are sometimes used where handrails end at a wall, such as on a balcony.

The height of the starting newel must be determined so that the newel can be cut to the correct length. The handrail should be centered on the newel's landing or flat surface (**Figure 10-76**). Use the TH to determine the height needed for the newel. Some carpenters set a string line representative of the handrail to find the correct newel height.

The starting newel has to be notched to fit over the starter step (**Figure 10-77**). Keep the center of the newel in line with the baluster centerline. Laying it out

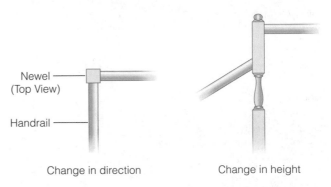

Change in direction   Change in height

FIGURE 10-74   Handrails change in both direction and height.

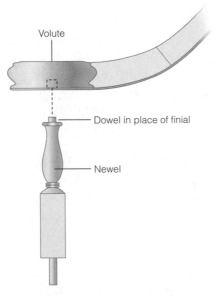

Volute

Dowel in place of finial

Newel

FIGURE 10-75   Some newels have a dowel pin in place of a finial, which accommodates a volute.

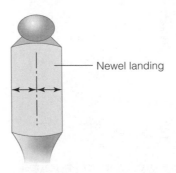

Newel landing

FIGURE 10-76   Handrail centered on newel landing

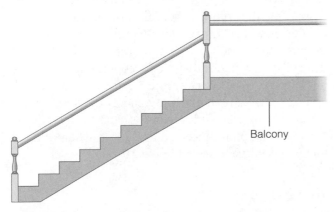

Balcony

FIGURE 10-73   A newel serving as a transitional point on a handrail that ascends to a horizontal one

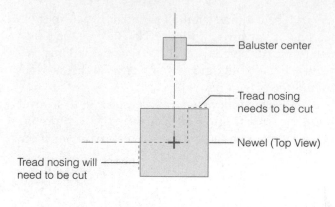

- Baluster center
- Tread nosing needs to be cut
- Newel (Top View)
- Tread nosing will need to be cut

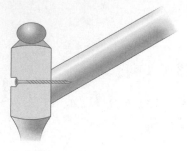

FIGURE 10-79   Handrail attached to newel landing

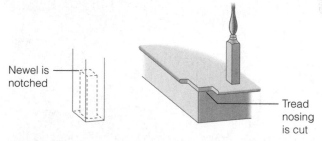

- Newel is notched
- Tread nosing is cut

FIGURE 10-77   Starting newel is sometimes notched to fit over the starter step.

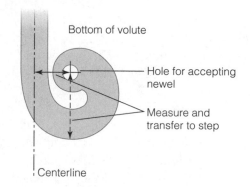

- Bottom of volute
- Hole for accepting newel
- Measure and transfer to step
- Centerline

- Outline of volute
- Transfer measurements from volute
- Centerline

FIGURE 10-80   Using the volute to determine placement of the newel

on the step will help you determine how much of the newel needs to be notched. Once notched, the newel is predrilled and fastened to the riser of the first step with lag bolts (**Figure 10-78**). Counterbore the predrilled holes so that wood plugs can be inserted to cover the screw heads. Make sure the newel is plumb; install shims if necessary.

The handrail is attached to the newel's landing, as shown in **Figure 10-79**. The handrail is cut with an angled cut and should meet the newel without any visible gaps. Predrill and counterbore the newel at the location of the handrail. Secure the two members with screws or lag bolts.

Volute-supporting newels are not in line with the centerline. The volute is in line with the centerline and then curves out to the left or right and ends above the newel. Use the volute to help find the proper position of the newel (**Figure 10-80**) by measuring the distance from the volute's centerline to the factory-drilled hole (which is there for accepting the pin at the top of the newel). Determine where the volute will end, and drill a hole into the tread of the starter step. Install the newel and further secure it by adding screws at the bottom as well as the top where the volute will eventually rest (**Figure 10-81**).

Rosettes

Rosettes are round or oval in shape and are used where handrails end at a wall (**Figure 10-82**). They

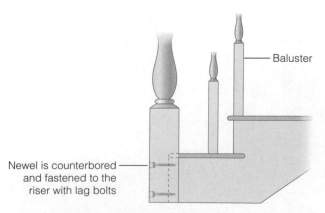

- Baluster
- Newel is counterbored and fastened to the riser with lag bolts

FIGURE 10-78   The newel is predrilled and fastened to the riser with lag bolts.

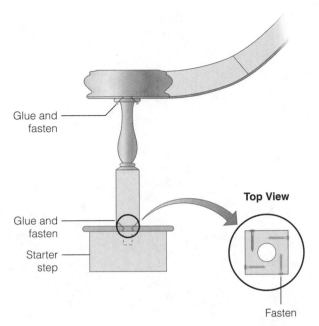

Glue and fasten

Top View

Glue and fasten

Starter step

Fasten

FIGURE 10-81   Secure the newel at top and bottom.

Volutes

A volute joins to a handrail and serves as a beginning point for the balustrade (**Figure 10-83**). Turnouts and easing members attach to the handrail and also serve as beginning and ending points (**Figure 10-84**). A properly attached volute should be level when attached to the sloping handrail (**Figure 10-85**).

To properly attach a volute, cut a pattern with the angle of the staircase (**Figure 10-86**). If the staircase is on a 35-degree angle, cut the pattern with what is called a long 35 or back-cut 35-degree angle. To get the back cut of a 35-degree angle, simply subtract 35 from 90 degrees, which leaves 55 degrees (the same pattern used as a guide for drilling baluster holes into the handrail). Cut the pattern at 55 degrees. With the volute and the pattern situated on a flat surface, scribe the volute where contact is made with the pattern (**Figure 10-87**). Next, the pattern is turned up, and then the volute is scribed (**Figure 10-88**). The volute is cut along the line (**Figure 10-89**). The handrail is cut square where it joins to the volute.

The volute is joined to the handrail with a double-ended screw. Predrill both the handrail and the volute with a drill bit that is slightly smaller than the screw.

are also used as ending points for balcony railings. Make sure rosettes are fastened to solid framing. Rosettes should not be fastened to sheetrock alone; that would be unsafe and could cause a fall.

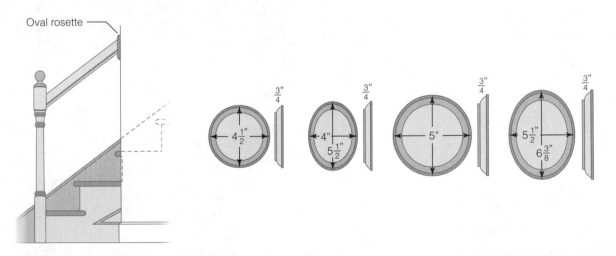

Oval rosette

$4\frac{1}{2}$" $\frac{3}{4}$"

4" $\frac{3}{4}$"
$5\frac{1}{2}$"

5" $\frac{3}{4}$"

$5\frac{1}{2}$" $\frac{3}{4}$"
$6\frac{3}{8}$"

FIGURE 10-82   Different styles and dimensions of rosettes

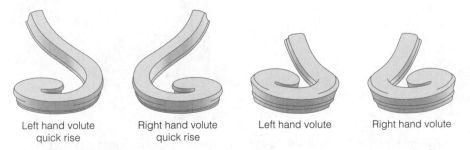

Left hand volute quick rise

Right hand volute quick rise

Left hand volute

Right hand volute

FIGURE 10-83   A volute

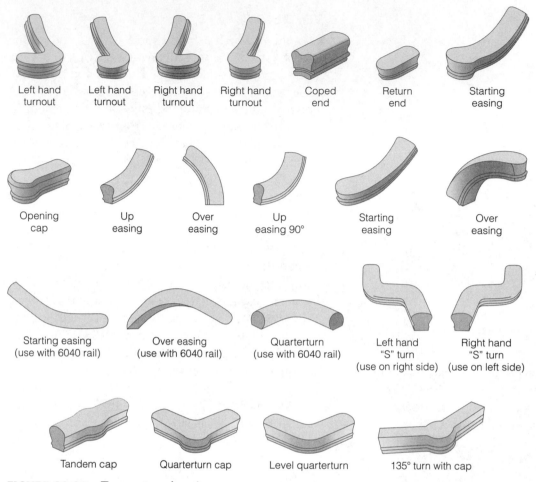

Left hand turnout

Left hand turnout

Right hand turnout

Right hand turnout

Coped end

Return end

Starting easing

Opening cap

Up easing

Over easing

Up easing 90°

Starting easing

Over easing

Starting easing (use with 6040 rail)

Over easing (use with 6040 rail)

Quarterturn (use with 6040 rail)

Left hand "S" turn (use on right side)

Right hand "S" turn (use on left side)

Tandem cap

Quarterturn cap

Level quarterturn

135° turn with cap

FIGURE 10-84 Turnout and easing

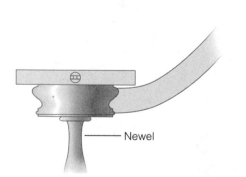

FIGURE 10-85 A volute is level when properly attached to handrail.

Newel

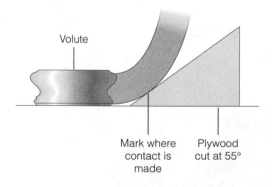

Volute

Mark where contact is made

Plywood cut at 55°

Bottom View of Volute

Square across line where contact was made

FIGURE 10-87 Scribe the volute where contact is made with the pattern.

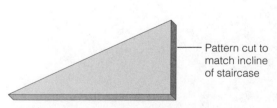

Pattern cut to match incline of staircase

FIGURE 10-86 A pattern cut with the angle of the staircase

Cut a ¼-inch section of the handrail to use as a guide for marking both pieces before drilling, so that the location of the pilot hole will line up correctly (**Figure 10-90**). Use wood glue on joining surfaces. The pieces should join without any gaps. Use fine-grit sandpaper to smooth the joints.

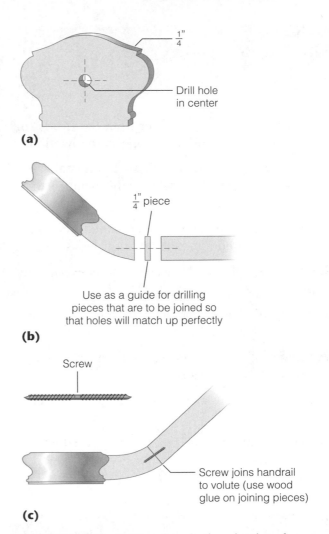

(a)

(b)

(c)

FIGURE 10-90 Drilling handrail and volute for double-ended screw

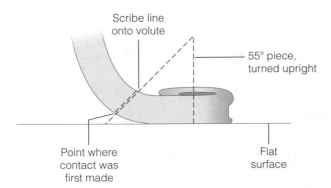

FIGURE 10-88 Turn the pattern upright and scribe the angle onto the newel.

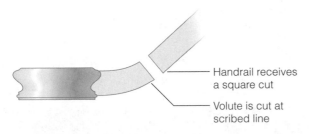

FIGURE 10-89 Cutting the volute

SUMMARY

- Each staircase presents a different set of challenges to overcome.
- The preferred angle of a staircase is between 30 and 35 degrees.
- Building codes regarding staircases may vary from state to state.
- There should be no less than 3 feet of clearance between handrails.
- The exact angle of the staircase must be determined before the staircase can be trimmed properly.

- A perfect stringer has equal risers, since risers of unequal height can cause accidents.
- Adding the width of one tread with one riser should produce a measurement of 17 to 18 inches.
- The number of risers will always be one more than the number of treads.
- When completing cuts for stringers with a jigsaw, avoid overcutting.
- The preferred spacing for balusters is no more than 4 inches apart.

KEY TERMS

balusters Member that extend from the treads to the bottom of the handrail; supports the handrail.

balustrade The upper members of the staircase; handrail, newels, and balusters are the members that make up the balustrade.

buttress cap Decorative cap that sits on the knee wall; accepts the bottom portion of the balusters.

fillets Pieces of wood inserted in between the balusters.

newels Sturdy beginning and ending points for the handrail and balustrade.

preacher An aid in marking tread and riser locations onto the finished stringer.

risers Vertical stair members, located in between treads.

rosettes Round or oval pieces that are used where handrails end at a wall.

stringer Trim piece made of 2-inch by 12 inch material; supports the treads and risers.

total rise Total of all the risers; the distance between finished floors.

total run Total of all the treads; the total horizontal distance of the staircase.

treads Horizontal members that are actually stepped on as a person ascends or descends a staircase.

unit rise The height of one riser.

unit run Distance of the tread, as measured from the face of one riser to the next (excluding the nosing).

volute Joins to handrail and serves as a beginning point for the balustrade.

REVIEW QUESTIONS

1. What is the preferred angle of a staircase?
2. Do all staircases ascend at the same angle?
3. What is the preferred height for a handrail?
4. Can a staircase be trimmed without knowing the exact angle of incline?
5. What are two types of stringers?
6. Can a staircase be both open and closed string?
7. If there are 15 treads, how many risers will there be?
8. What are two things that can be done by using a temporary handrail?
9. Describe what is known as the total rise.
10. Does the total rise affect the height of the risers?
11. When calculating riser height, the first thing you do is to divide the total rise by what number?
12. What is a characteristic of a mitered stringer?
13. What should you get when you add the width of one riser and one tread?
14. The height of a balcony railing should be no less than what?
15. What could risers of unequal height lead to?
16. What are fillets, and what are they used for?
17. What is the back cut or long 35-degree angle?

Columns

CHAPTER 11

OBJECTIVES

After studying this chapter, you should be able to

- Determine the height and width of a column
- Cut and join the sides of a column
- Lay out a column for trim and fluting pattern
- Flute and add trim to a column
- Install columns

INTRODUCTION

Columns differ in size and shape, but usually they are square or round (**Figure 11-1**). Their surface can be smooth or fluted with shallow vertical grooves. The top of a column can be fitted with a cap or trimmed with crown, and the bottom can be fitted with a base molding. Beneath the cap there can be a subtrim, usually made from a molding of a smaller dimension.

Columns can go from floor to ceiling, or they can go from a half wall to the ceiling or to the underside of a beam or header. Columns vary in height and width. The number of flutes per side and the spacing of the flutes also can vary.

Joining edges of a square column are cut at 45 degrees (**Figure 11-2**), which creates a joint that is suitable for painting or staining. Square-butted edges would not be suitable for painting or staining due to the change in grain pattern, which would affect the appearance by showing obvious differences in paint absorption.

Columns made of composite materials can be purchased at building supply centers or online companies. Composite columns (fiberglass, fiberglass-reinforced gypsum, and polymer) simply need to be installed. They can also be purchased with a matching **capital** and base. Columns can be constructed by the finish carpenter as described in the following sections. ■

(a)

FIGURE 11-2   Edges of a column join with a 45-degree cut.

(b)

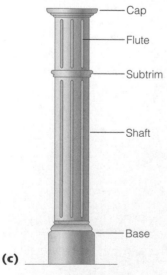

Cap

Flute

Subtrim

Shaft

Base

(c)

FIGURE 11-1   Columns

TYPES OF COLUMNS

Columns are either structural or nonstructural. Columns made by the finish carpenter are typically nonstructural, meaning they do not support a substantial amount of weight. Load-bearing columns are usually engineered and designed for load-bearing purposes. Some load-bearing columns are made by wrapping a 4-inch by 4-inch or a 6-inch by 6-inch timber with ¾-inch material, and then adding trim (**Figure 11-3**). Columns are also formed by wrapping rough framing with sheetrock.

Pilasters, unlike true columns, have only three sides instead of four. Made of 1-inch by 6-inch or 1-inch by 4-inch material, only the face side is fluted. Pilasters are used to break up rooms that share one continuous wall, such as between a living room and

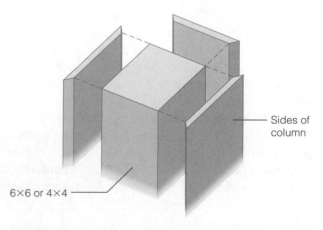

Sides of column

6×6 or 4×4

FIGURE 11-3   Some columns are made by wrapping a 4-inch by 4-inch or 6-inch by 6-inch piece of material suitable for finishing.

dining room. The principles of making a pilaster are the same as those of making a column.

MATERIALS

Columns constructed of fiberglass, fiberglass-reinforced gypsum, and PVC can be purchased at building supply centers or online companies; these products save time and are easy to install. Composite columns are available in different styles (round, tapered, fluted, etc.) and they can be fitted with a matching capital and base. Columns created by the finish carpenter are constructed using wood or a wood product. A solid wood is ideal for making a column because the edges, after joining the sides, can be sanded or slightly rounded over using a router. Plywood is made of different layers, and the finished layer is easily damaged or peeled away at edges where the sides of a column join (**Figure 11-4**). Particleboard and fibrous manmade products are also easily damaged or chipped away at the

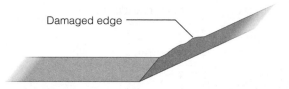

FIGURE 11-4 Plywood is made up of layers and can be easily damaged at corners.

edges, which affects the look of the finished product. Solid wood holds sharp edges better and can be sanded smooth without causing further damage to the column. Solid wood also can be fluted, whereas plywood cannot be fluted without damaging the finished layer and exposing the filler layers beneath.

In short, the advantages of using solid wood include

- Edges are less likely to become damaged.
- Edges can be shaped or sanded.
- Solid wood can be fluted.

 **PROCEDURES: MAKE AND INSTALL A COLUMN**

 SAFETY TIPS

- Wear OSHA-approved safety glasses, footwear, and hard hat.
- Wear proper clothing. Baggy clothing and unbuttoned cuffs can get caught in power equipment.

6. Add trim.
7. Install the column.

These steps are explained in detail in the sections that follow.

NOTE

Determining the height and width of the column (as well as the trim sizes) is done first, since that information is necessary in laying out the fluting pattern so that it will be uniform and evenly spaced. ∎

11-1 PREPARING TO MAKE A COLUMN

A logical order of steps in constructing a column would be as follows:

1. Determine the height of the column.
2. Determine the width of the column.
3. Rip the sides of the column on a table saw.
4. Join the sides.
5. Flute the column.

11-2 DETERMINING THE HEIGHT OF A COLUMN

The height of a column is determined by taking a vertical measurement at the location where the column is going to be installed. If there are to be several

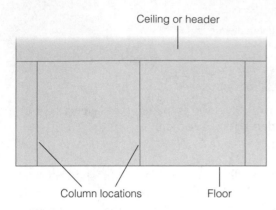

FIGURE 11-5 Take measurements at each column location.

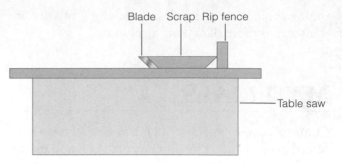

FIGURE 11-6   Use a scrap piece to test the blade setting.

columns, each location should be measured to ensure an accurate fit of each one. An unlevel floor or ceiling can cause a difference in each measurement (**Figure 11-5**).

11-3 DETERMINING THE WIDTH OF A COLUMN

The width of a column is usually predetermined by architectural drawings or sketches. There is no standard or predetermined width for columns; it is a matter of aesthetics, or deciding what width will look best for the given situation. Less width gives a column a narrow, slender look, just as more width gives it a stocky or bulky look. The extreme in either direction will result in a column that looks out of place compared to its surroundings.

11-4 RIPPING THE SIDES OF A COLUMN

Ripping the sides of the column can be done with a table saw set on 45 degrees. Cut a test piece to check the accuracy of the blade and to make sure the fence is set the proper distance away from the blade. The stock used should be as straight as possible; when you can, you should true it on a jointer before ripping it on the table saw. Hold material securely against the fence and flat on the saw base when ripping. **Featherboards** will help hold material securely to the saw base while ripping the material. Featherboards, simple boards with a series of 45-degree cuts along their edge, are

designed to keep material pressed firmly against the saw base and rip fence.

Begin by picking the straightest material available. True the material on a jointer if possible. Straight material will ensure a column with tight joints. A sharp saw blade will help to achieve tighter joints as well.

Set the angle of the table saw blade to 45 degrees. Adjust the table saw fence to the width needed for the sides of the column. Test for proper blade angle and check the rip fence setting by ripping a piece of scrap material on the table saw (**Figure 11-6**). Adjust the rip fence setting if necessary. To further check the angle of the blade for accuracy, rip a piece of scrap and cut four 1-inch pieces, and then fit the pieces together in the form of a square. After squaring the pieces with a speed square, examine the joining corners, which should fit together without any gaps (**Figure 11-7**).

Once the saw is set properly, all the sides for the column can be ripped. Use stock that is a little longer than the finished product will be so that the ends can be squared off later, after the four sides have been joined. When you're ripping material on the table saw, the material should be kept tight against the rip fence at all times and should be kept flat on the saw base as

FIGURE 11-7   Check the test pieces to see if the corners join properly.

well. Failure to do so will result in corners that do not join properly.

When ripping the sides of a column, remember the following:

- Use straight material.

- True the material on a jointer.

- Set the blade angle to 45 degrees.

- Set the rip fence of the saw to the desired cutting width.

- Test both settings (rip width and blade angle) by ripping a piece of scrap material.

- Hold the material securely against the rip fence and flat on the saw base when ripping.

CAUTION

Before making cuts, inspect material to make sure there are no hidden nails or staples present.

11-5 JOINING THE SIDES OF A COLUMN

Joining the sides of a column is done with a jig (which holds the pieces together properly), wood glue, and finish nails or wood screws (**Figure 11-8**). A ribbon of

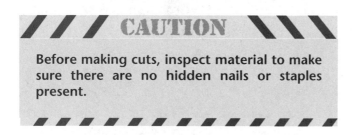

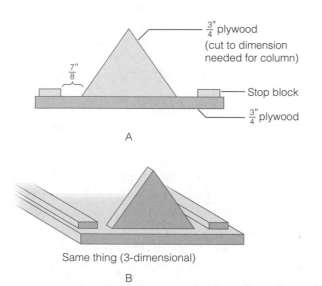

FIGURE 11-8 Joining the sides of a column is done with a jig, wood glue, and finish nails or trim screws.

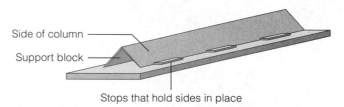

FIGURE 11-9 A jig helps hold the pieces while nailing.

wood glue is run the length of each joint. Finish nails are driven in alternating patterns that hold the sides together with a stitching effect. Trim screws are added for extra support and are used to close small gaps where the corners join.

A simple jig helps when assembling the sides of the column but is not necessary. A jig helps hold the sidepieces in a proper position while nailing (**Figure 11-9**). Apply a ribbon of wood glue before joining the pieces. Start cross-nailing at one end and work toward the other end. Hand pressure is often enough to close any gaps between the two pieces. Sometimes the pieces will require slight manipulation to keep the corners properly joined. Stay as far away from the edge as possible with nails and screws, as there is less wood closer to the edge for nails and screws to hold onto (**Figure 11-10**).

Cut and install support blocks as shown in **Figure 11-11**. Support blocks serve three main functions:

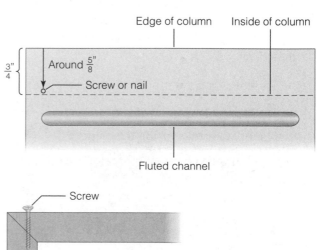

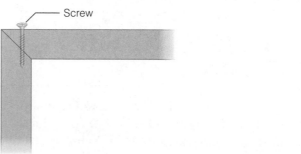

FIGURE 11-10 Keep away from the edge with nails and screws.

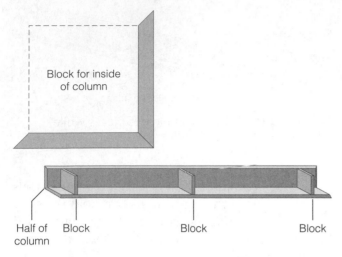

FIGURE 11-11 Support blocks add strength to a column.

(a)

to help keep the column square, to give the four sides something solid to attach to, and *to prevent corners from separating.* Cutting the blocks ¹⁄₁₆ inch less than the inside measurement of the column will help to further ensure tight joints. Support blocks help create a more stable column. After the column has been fluted, trim screws can be inserted and countersunk at block locations.

First assemble the two halves, and then the two halves can be joined together (**Figure 11-12**). One end of the column can be cut off squarely. A measurement can then be pulled from the square end, and an accurate length can be marked where another square cut is required. The column should be cut to its final length before marking the flute layouts.

When joining the sides of a column, remember the following:

- A jig is useful when assembling the sides of a column.

- Use wood glue on the joints.

- Install support blocks inside the column.

- Cut the column to final length before marking the flute layouts.

11-6 FLUTING THE COLUMN

Fluting the column is done with a router equipped with a fluting bit (**Figure 11-13**). After the trim locations have been marked, layouts indicating individual flute locations can be made. Test the fluting bit on a piece of scrap to ensure proper depth, and make adjustments if necessary before routing the finished

(b)

FIGURE 11-12 After the two halves have been assembled, they can be joined to form one unit.

column. Fluting generally stops at a planned distance away from trim locations.

Mark locations where there is going to be any trim (cap, base, subtrim, etc.). Knowing the trim locations means that the fluting can be stopped at a planned and uniform distance away from the trim. Use a square to continue the trim layout lines completely around the column. Measure and mark a line that will represent a stopping point for the fluting. Writing the words *space* and *trim* between the appropriate lines will help by serving as reminders of where to stop when the actual routing begins (**Figure 11-14**). The lines will also be helpful later on while nailing the trim squarely onto the column.

Knowing the width of the flute is necessary when marking the flute layouts onto the column. The depth

(a)

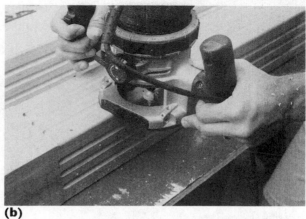

(b)

FIGURE 11-13   Fluting is done with a router equipped with a fluting bit.

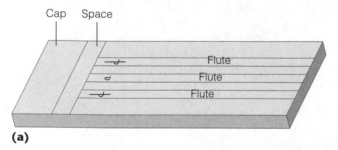

(a)

(b)

FIGURE 11-14   Marking trim locations on column so that fluted channels can be made using a router

FIGURE 11-15   Using a combination square to lay out fluting pattern

setting of the fluting bit affects the width of the flute. For evenly spaced flutes, the width must be known so that the layouts can be spaced evenly. Adjust the bit and test it on a piece of scrap material; afterward, the width of the channel can be measured and the measurement can be factored into the spacing equation. The column can then be laid out using a combination square (**Figure 11-15**). Double-check to make sure the layouts are spaced evenly before using the router.

When using the router, let the rpm build up before letting the bit make contact with the wood. Stop

the flutes at the layout lines made earlier. When the router fence is set up to rout the center flute, all four sides can be done, and likewise for the outside flutes. An entire column can be fluted with just two different settings of the router fence (**Figure 11-16**).

To sand the fluted channels (**Figure 11-17**), wrap sandpaper around something that has the same radius as the flute channel (a wooden dowel, a drill bit, a socket from a ratchet set, etc.). Once the flutes have been completed, the column can be trimmed and the screws can be inserted into the support blocks.

Remember the following when fluting a column:

- Mark trim and space locations.
- Space the flute layouts equally.
- Stop the flute channel a planned distance away from trim locations.
- Sand the fluted channels.
- Add trim.

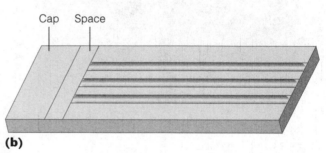

CAUTION

Make sure the rpm is built up before blades make contact with stock. Hold tools firmly and in a safe manner.

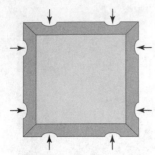

A. Locations that can be cut with one setting of the router fence

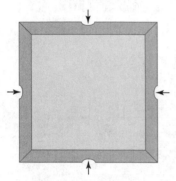

B. Locations that can be cut with one adjustment of the router fence

FIGURE 11-16   An entire column can be fluted with two adjustments of the router fence.

FIGURE 11-17   Sanding the channel

11-7 ADDING TRIM TO THE COLUMN

Adding trim to the column can be done after the fluting is complete (**Figure 11-18**). Trim layout lines help to ensure that the trim is being put on squarely. Some carpenters prefer to install the cap, or crown, after the

FIGURE 11-18   Trim is added to the column after the fluting has been completed.

column has been installed so that there is minimal gap where the column meets the ceiling. Trim is fastened to the column using wood glue and finish nails.

11-8 INSTALLING THE COLUMN

Installation of the column is done using a level to ensure that the column is plumb. A column should be plumbed in both directions (**Figure 11-19**). Once the column is plumb, scribe reference marks on the floor and ceiling to help gauge any movement that may occur during installation (e.g., while finish nails or

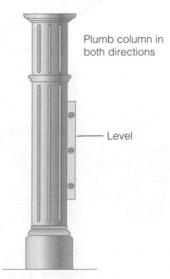

Plumb column in both directions

Level

FIGURE 11-19   Check for plumb in both directions when installing a column.

trim screws are being added). Secure the column with finish nails or trim screws (**Figure 11-20**). **Figure 11-21** shows the top and bottom of a column after it was finished. Some carpenters prefer to install crown around the top of the column after it has been secured in place.

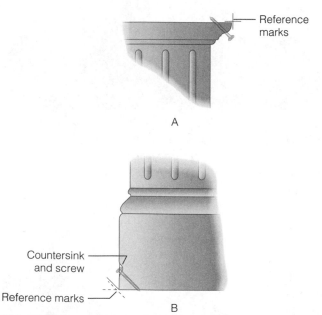

FIGURE 11-20   Secure the column with finish nails or trim screws.

(a)

(b)

FIGURE 11-21   Top and bottom of the finished column

SUMMARY

- Columns differ in size and shape.
- The height and width of the column must first be known before anything else can be done.
- Knowing the height and width of the column and the trim sizes that are going to be used is necessary so

that the fluting pattern can be laid out in such a way that everything will be uniform and evenly spaced.
- When you are ripping stock on the table saw at 45 degrees, the stock must remain flat on the saw base and be held securely against the fence.

- Making a test rout is necessary in order to establish equal spacing of the flutes.
- Support blocks inside the column will ensure a firm, long-lasting product.
- The column can be cut to length after all four sides have been joined.

- All routing must take place before any trim can be nailed onto the column.
- Even with many individual flutes on a column, the guide on the router only needs to be changed once after the initial setup.
- When you are installing a column, it must be plumbed in both directions.

KEY TERMS

capital The top, or cap, of a column.

featherboard A board with a series of 45-degree cuts along its edge, designed to hold constant pressure on material as it is being ripped.

pilaster A three-sided, shallow column usually installed on a wall to divide rooms (such as between a kitchen and dining area, or between a dining and living area).

REVIEW QUESTIONS

1. What is the first step in making a column?
2. Why is knowing the width of the flutes important?
3. When ripping material on the table saw at 45 degrees, why is it important to run a test piece through the saw?
4. Is it necessary to make and use a jig when making columns?
5. What has to be done before the lengths of the flutes can be determined?
6. What can cutting samples of the trim you are using help you with?
7. Is it safe to say the layout of the flutes and the trim you are using go hand in hand?
8. What is one benefit of squaring trim lines, spacing lines, and flute layout lines completely around the column?
9. Is it absolutely necessary to square trim, spacing, and flute layout lines completely around the column, on all four sides?
10. When installing a column, is it necessary to plumb it in both directions?
11. How many blocks cut to fit inside the column were sufficient for the column in the example discussed in this chapter?
12. Name one purpose the support blocks serve.
13. Briefly explain how to sand the fluted channel.
14. Why is it necessary to complete all the routing before adding any trim?

Cabinet Installation

OBJECTIVES

After studying this chapter, you should be able to

- Install upper cabinets that are level and plumb
- Install base cabinets that are level and plumb
- Cut and install countertops
- Make a jig for attaching drawers and doorknobs and pulls
- Attach doors to cabinets and mount pulls

INTRODUCTION

Most cabinets are made or purchased at one location and are then delivered to the location where they are to be installed. When installing cabinets, the installer has to contend with floors and walls being out of level and out of plumb, as well as being out of square. Cabinets are square, and they have to be installed in such a way that they are level and plumb, despite the condition of the floor and walls.

Once the cabinets have been shimmed and are set properly, the installer must close any gaps that may exist between the unit and the wallboard. Gaps are closed with screen molding or something similar. Screen molding is made of the same material as the cabinets and is thickness-planed to a dimension of ¾ inch by ¼ inch. Where the cabinets meet the floor, a facer board sometimes has to be attached to the front of the toe kicks in order to conceal a gap.

This chapter covers typical cabinet installation procedures and discusses the most commonly used methods of installation. It also addresses some of the problems finish carpenters face in installing cabinets, such as those mentioned in this introduction. Installation methods are basically the same despite the materials used for the construction of the cabinets. ■

TYPES OF CABINETS

Figure 12-1 shows some typical cabinets, both upper and lower (or base) cabinets. These were made by a custom cabinet company and then delivered to a new home under construction. Cabinets can also be constructed with a material known as melamine.

FIGURE 12-1 Cabinets

MATERIALS

Cabinet stiles and rails are usually made of a solid wood. Doors and drawer fronts are also made of solid wood so that their edges can be shaped.

The sides of the cabinet as well as the shelves are typically made from plywood (the same species used for stiles and rails). In some cases, where cabinets are going to be painted, cabinetmakers may use MDF for the sides and shelves, though they still use a solid wood for the stiles and rails.

Cabinets are also constructed with a material known as melamine. It is a durable material that at present is being used mostly for commercial cabinetry. Steel or metal cabinets are used in some cases.

Countertops, if they are going to be covered with tile or plastic laminate, can be of plywood, MDF, or particleboard. Solid-surface countertops are now being widely used. Countertops can be made of stone, granite, concrete, or Corian.

PROCEDURES: INSTALL CABINETS

Despite the many different types and styles of cabinets, the installation process is basically the same.

SAFETY TIPS

- Make sure ladders are secure before climbing up them.
- Check safety features on tools to make sure they are functioning properly.
- Hold tools firmly and in a safe manner.
- Always wear safety glasses when operating power tools.

12-1 INSTALLING WALL AND BASE CABINETS

The wall cabinets, sometimes referred to as **uppers**, are installed 18 inches above the **lowers**, or base, cabinets (**Figure 12-2**). Many finish carpenters find that

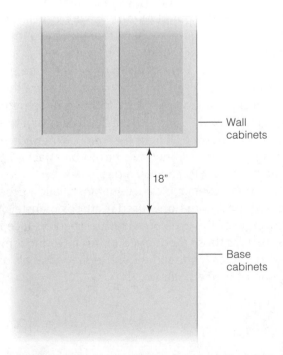

Wall cabinets

18"

Base cabinets

FIGURE 12-2 Wall cabinets are installed 18 inches above the base cabinets.

setting the upper cabinets first is easier and more convenient. Uppers are easier to set when base cabinets are not in the way. However, some carpenters prefer to set the base cabinets first. As long as both upper and lower cabinets are level and plumb and are spaced an equal distance apart, it does not make a difference as to which is set first.

Installing Wall Cabinets

The first thing to do is to find the highest part of the room (**Figure 12-3**). Use a standard level or a laser level to do this. Once the highest part of the floor has been found, measure up and make a mark on the wall

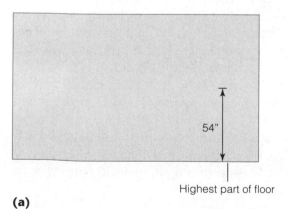

Level

FIGURE 12-3 Locating the highest section of floor

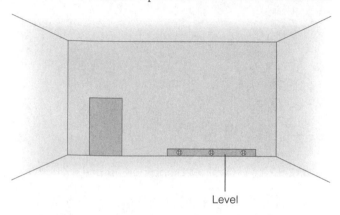

54"

Highest part of floor

(a)

Continuous level line

54"

(b)

FIGURE 12-4 Make a level line on the wall as a reference for upper cabinets.

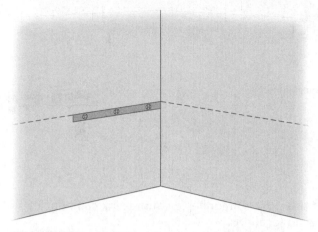

FIGURE 12-5 Using a level to mark lines

at 54 inches, which is 4 feet 6 inches. This mark represents the bottom of the upper cabinets. Continue making marks at 54 inches, and then make a continuous level line on the walls where the upper cabinets are to be placed (**Figure 12-4**); or you can snap chalk lines on the walls where the uppers will be placed (**Figure 12-5**). A laser level is ideal for marking this 54-inch line. Double-check to make sure that lines are level.

Some carpenters find it easier to install wall cabinets if they mount a support board on the wall at the height of the chalk line (**Figure 12-6**). This gives the cabinets something to rest on while being attached. The painter can repair the holes in the sheetrock later. Locate the studs in the wall and transfer those measurements to the cleats of the cabinets. Then predrill and countersink the cabinet cleats at stud locations (**Figure 12-7**).

Above the sink area, the uppers need to be laid out in such a way that the sink will be centered between them (**Figure 12-8**). Usually there is a window above the kitchen sink. If this is the case, make sure the uppers are set an equal distance on either side of the

Side View

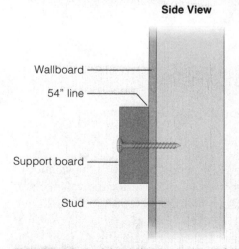

Wallboard

54" line

Support board

Stud

FIGURE 12-6 Mounting a support board at level line

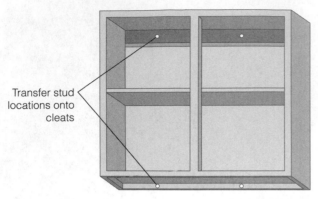

FIGURE 12-7 Predrill and countersink cabinet cleats at stud locations.

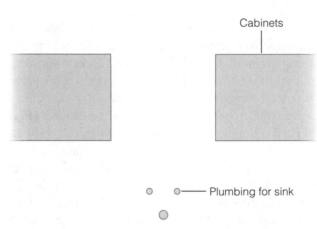

FIGURE 12-8 Wall cabinets need to be laid out so that the sink is centered between them.

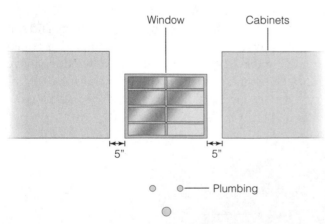

FIGURE 12-9 Equal distance on either side of the window

opening (**Figure 12-9**), and do this with the valance in mind. Enough room has to be left for the valance and any appliances (such as a range vent hood), as shown in **Figure 12-10**.

Some installers fasten all consecutive upper cabinets together and install them as one unit. If all the

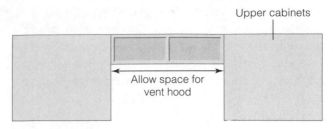

FIGURE 12-10 Smaller cabinet between uppers

FIGURE 12-11 Cabinet being held in place on level line

uppers are joined before installation, a cabinet lift is useful. The lift has wheels so that the cabinet can be pushed right up against the wall. If a cabinet lift is not being used, several people can hold the cabinets against the level line (**Figure 12-11**).

With the cabinets being held in place, drive 3-inch screws through the cleats and into the studs, using a drill. (A cordless drill is most convenient because there is no cord to deal with.) Check the cabinets for plumb, and shim them if necessary. Once they are plumb, finish screwing the top and bottom cleats completely.

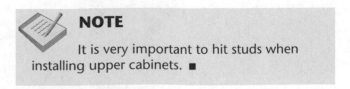

NOTE

It is very important to hit studs when installing upper cabinets. ■

Finish the installation by closing any gaps with molding or shoe mold (**Figure 12-12**) or, if only minor gaps are present, by shaving the cabinet to fit. Sometimes the sheetrock can be cut enough so that the cabinets can be fitted without leaving gaps. Most professional cabinet installers use **screen molding**

(a)

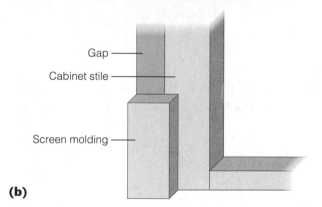

(b)

FIGURE 12-12 Screen molding covers gap between cabinet and wall.

FIGURE 12-13 Leveling base cabinets

FIGURE 12-14 Checking level from front to back

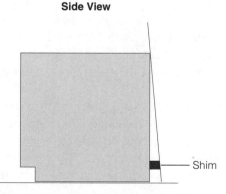

FIGURE 12-15 Shimming base cabinet from the back so that the unit will be level

and crown to conceal the gaps that are left behind. Screen molding is usually ¼-inch by ¾-inch and is used mainly to cover the edges of veneer (which shelves are usually made of).

Installing Base Cabinets

To install a base cabinet, set the cabinet against the wall and place a level on the top (**Figure 12-13**), or measure down 18 inches from the upper cabinets and snap a continuous level line. (The finished height of base cabinets is usually 36 inches.) Check for level from front to back on lower cabinets (**Figure 12-14**). The line on the wall can be a reference for checking level in that direction.

Figures 12-15 and **12-16** demonstrate the shimming and adjustment procedures for leveling the base cabinets. First, shim where necessary to get the cabinet in line with the level reference line. Once this has

Front View

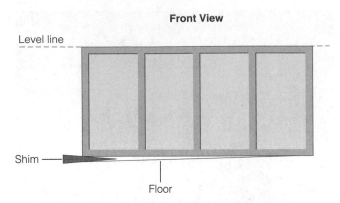

FIGURE 12-16 Shimming base cabinet at the floor so that the unit will be level and plumb

(a)

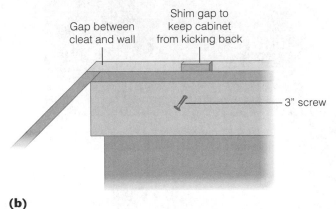

(b)

FIGURE 12-17 Placing a screw through the cleat of a base cabinet after it has been leveled

been done, it is just a matter of leveling the unit from front to back.

Once the cabinet is in line with the reference line and is good from front to back, countersink the screw rail at stud locations. (This does not need to be done at every stud location; every other stud should be more than sufficient.) Use 3-inch screws to fasten cabinets to the wall (**Figure 12-17**). If there is a space between the wall and the screw rail due to the walls being out of plumb, insert a shim to fill in the gap. Doing this will prevent the cabinet from kicking back (which means that it would no longer be level).

If two separate cabinets are sitting next to one another, their fronts can be flushed and a couple of screws can be placed through the face frames after countersinking. Screws can be hidden by placing them near a hinge area or in a drawer opening (**Figure 12-18**). Another option is to place a 1¼-inch screw inside the cabinet at the top and bottom, as well as the front and back of the unit (**Figure 12-19**). Holes can be puttied with a stain- or paint-grade wood filler.

Install all base cabinets and vanities in this manner. The height of the vanity may be less than that of the kitchen cabinets. In this case, just adjust the line according to the height of the vanity. Usually, vanities are at a height of 32½ inches without the countertop.

Most of the time, the toe kicks of base cabinets will need to have a facer board nailed over them. This dresses up the bottom of the cabinets and closes any gaps, such as where two cabinets meet in a corner (**Figure 12-20**). Measure and cut the facer board to the size needed. Flooring will usually come up enough to cover small gaps. Shoe mold is sometimes

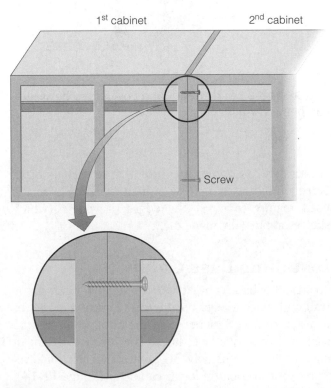

FIGURE 12-18 Screws joining face frames can be hidden from view.

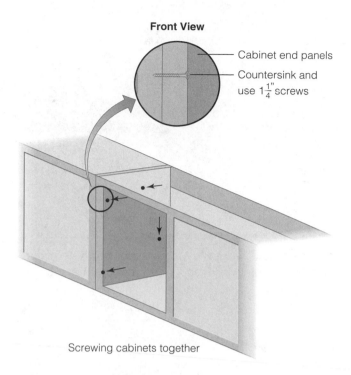

Front View

Cabinet end panels

Countersink and use $1\frac{1}{4}''$ screws

Screwing cabinets together

FIGURE 12-19   Placing screws inside unit

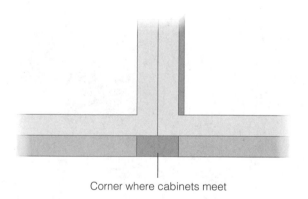

Corner where cabinets meet

FIGURE 12-20   Facer board nailed over toe kicks

added after the flooring has been installed to close gaps between the toe kicks and the finished flooring (**Figure 12-21**).

> ### NOTE
>
> Installing the cabinets without the doors and drawers in them makes the task easier. This is an especially good idea when you consider the fact that the units may have to be moved in and out of the area several times before you finish the installation. ■

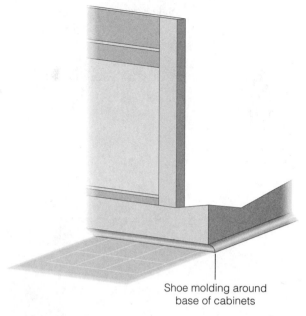

Shoe molding around base of cabinets

FIGURE 12-21   Shoe molding around base cabinets

12-2 INSTALLING COUNTERTOPS

To attach countertops, simply position them on top of the cabinets and secure them with screws. In some cases, it may be necessary to **scribe**, or mark, the countertop to fit the wall; to scribe the countertop, use a gauge block or a compass (something that will maintain a consistent distance). Scribing the countertop will allow it to fit tightly with the contour of the wall. If possible, insert screws from the bottom side so that the countertop can be removed if necessary. If you are using a countertop that has a warranty, the warranty may be void if the countertop is not installed by the manufacturer. Corian is one such countertop.

Sometimes ¾-inch plywood is used for the countertop, as well as particle board and medium-density fiberboard (MDF), and is later covered with plastic laminate or tile. If plastic laminate is going to be used, a 1½-inch strip needs to be nailed on, flush with the top of the plywood (**Figure 12-22**). If tile is going to be used, glue and nail on stain-grade wood or something similar so the strip will be flush with the finished height of the tile (**Figure 12-23**).

Cutting Countertop for Sink

The sink being used should come with a template for cutting the hole for it. If there is no template, you will have to measure the sink and transfer those measurements to the

Side View

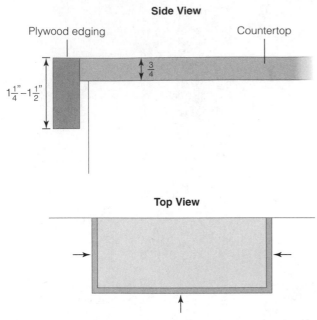

FIGURE 12-22 Edging for countertops

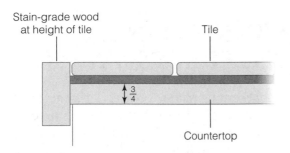

FIGURE 12-23 Stain-grade edging flush with tile

countertop. Solid-surface countertops are manufactured so that the sink hole is already present. Cutting a hole in granite or Corian requires specialty tools and may void the warranty if not performed by the manufacturer.

> **NOTE**
>
> When you are measuring for the sink, it is important to remember that the hole will need to be smaller than the flange around the sink. ■

Layout lines on the countertop should be square and should run parallel with the front of the counter (**Figure 12-24**). The cutout for the sink should be centered as shown in **Figure 12-25**. The sink should be centered with the window and with the doors of the base cabinet below.

Depending on the thickness of the backsplash, you may not be able to center the sink perfectly

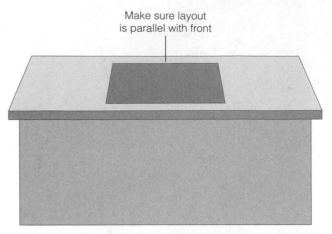

FIGURE 12-24 Layouts for the sink should be square and run parallel with the front of the countertop.

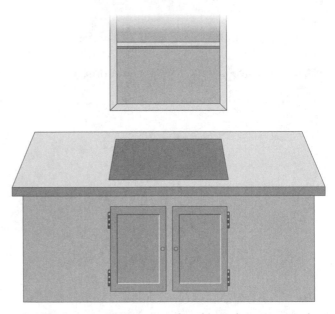

FIGURE 12-25 The cutout for the sink is centered with both window and cabinet doors below.

from front to back. Remember that the part of the sink nearest the front has to clear the cabinet rails below.

Once you have the template drawn onto the countertop, you can make the cut with either a jigsaw or a skill saw. Start by drilling a ½-inch hole big enough for inserting a jigsaw blade. Use the jigsaw to make the cutout (do not forget that the corners of most sinks are rounded over). If cutting a hole on a counter that already has plastic laminate on it, lay down blue painter's tape around the layout and use a fine-toothed blade so as not to damage the surface (**Figure 12-26**). If you are using a skill saw, a plunge cut will have to be made to begin each cut.

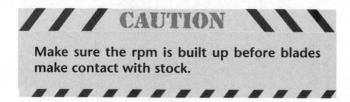

CAUTION

Make sure the rpm is built up before blades make contact with stock.

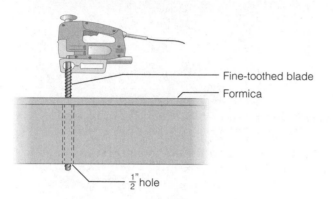

Fine-toothed blade

Formica

$\frac{1}{2}"$ hole

FIGURE 12-26 Using a fine-toothed blade to cut out a countertop covered with plastic laminate.

12-3 INSTALLING KNOBS AND PULLS

If the cabinets being installed have already been drilled for knobs and pulls, simply attach them with the proper hardware. If the holes have not yet been drilled, a jig can be used for making the holes. Using a jig will ensure that all the doorknobs will be consistent with one another. A template for marking the drawers can be used to make the task easier and faster.

NOTE

Building supply centers, online companies, and manufacturer supply catalogs have an endless selection of cabinet hardware to choose from (knobs, pulls, hinges, drawer guides, etc.). ■

After deciding where to put the knobs or pulls, use the intended hardware for making a template, or jig (**Figure 12-27**). Measure the screw holes from center to center. Find the center of the drawer in both directions and make a level line (a torpedo level works fine), or measure down an equal distance on

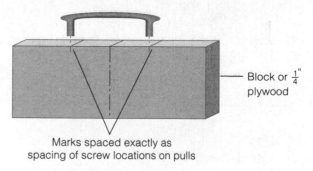

Block or $\frac{1}{4}"$ plywood

Marks spaced exactly as spacing of screw locations on pulls

FIGURE 12-27 Make a jig to drill for knobs and pulls.

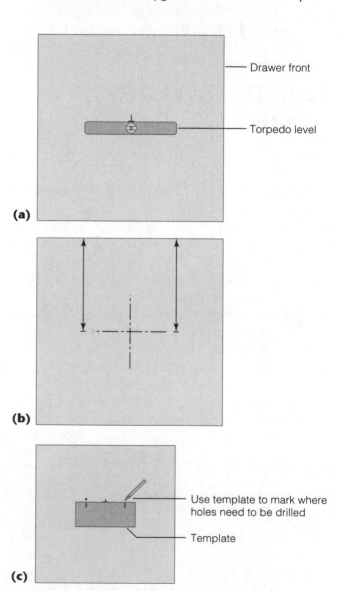

Drawer front

Torpedo level

(a)

(b)

Use template to mark where holes need to be drilled

Template

(c)

FIGURE 12-28 Marking layout onto drawer front

each side of the center mark (**Figure 12-28**). Use the template to mark the hole locations. Then you are free to drill through the drawer. A forstner bit may be used to drill-recess the holes inside the drawer. Doing this will allow the head of the screw to go in far

Top View

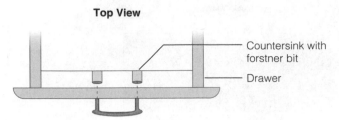

Countersink with forstner bit

Drawer

FIGURE 12-29   A recessed hole allows screw to be threaded into drawer pull.

enough so that the screw can be threaded into the pull (**Figure 12-29**).

12-4 INSTALLING DOORS

Mount the cabinet doors before installing the pulls. This makes for easier placement of the pulls from an aesthetic point of view. Before you mount a door, the hinges need to be mounted onto the door.

A good rule of thumb to remember for the placement of hinges is shown in **Figure 12-30**. This is a common practice, and one that is easy to remember. Simply use the hinge itself to mark the distance from either end, and then screw the hinges onto the doors. As you do this, remember which doors swing left and which ones swing right; this sometimes matters, especially when the pattern on the doors is designed to have a top and bottom (**Figure 12-31**).

Find out how much allowance the door has by measuring the door, and then measure the opening over which it is going. Doors are usually made at least

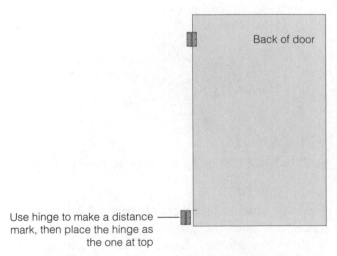

Back of door

Use hinge to make a distance mark, then place the hinge as the one at top

FIGURE 12-30   Use the hinge to mark location of hinges.

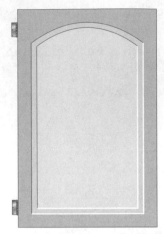

Be sure to place hinges on correct side of door; a door with a pattern such as the arch on this one is not reversible

FIGURE 12-31   A door with an arched design, with the arch designating the top

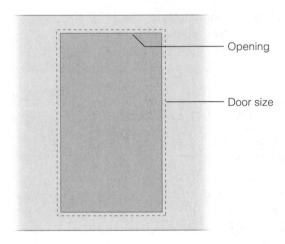

Opening

Door size

FIGURE 12-32   Doors are usually made at least ½ inch bigger than the opening.

½ inch bigger than the opening (**Figure 12-32**), but measure the door and the opening to make certain. If this is the case, clamp a straight 1 × at ¼ inch beneath the bottom opening (**Figure 12-33**). This will give the doors a shelf to rest on while you mark the hole pattern of the hinges onto the stile (**Figure 12-34**). Be sure to center the doors from left to right as well (**Figure 12-35**).

Once the holes have been marked, drill pilot holes with a bit that is slightly smaller than the hinge screws. Doing this will prevent any mishaps before they happen, such as a screw walking or moving slightly away from its intended mark, which would result in a screw that, once sunken, would appear crooked to the viewer. Drilling pilot holes will also prevent screws from splitting the wood. You can mount all doors for upper and lower cabinets in this fashion.

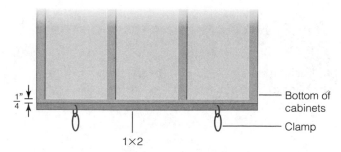

FIGURE 12-33 Clamp a level straight edge across the bottom of the cabinets.

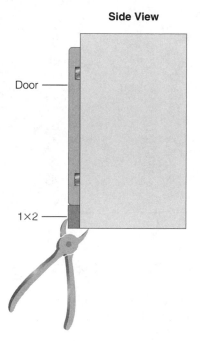

FIGURE 12-34 Doors resting on level straight edge

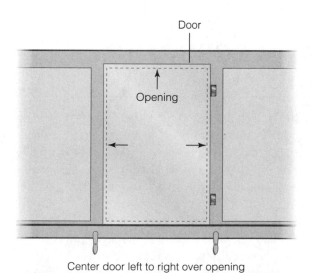

FIGURE 12-35 Make sure to center doors from left to right.

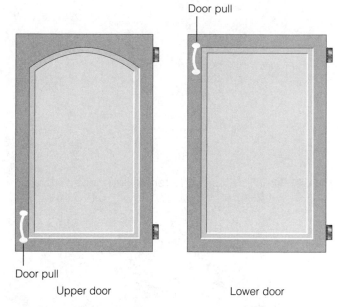

FIGURE 12-36 Where door pulls are most commonly mounted

Once all the doors have been installed, the door pulls can be mounted. **Figure 12-36** shows the location where door pulls are most commonly mounted.

> **NOTE**
>
> Most factory cabinets have had the hinges and doors installed at the factory. In some cases, they even have predrilled holes for the handles and pulls. ■

Mounting Door Pulls

A simple jig can be made for the purpose of mounting door pulls. Using the jig ensures that all pulls are consistent with one another. First, after deciding where on the door the pull looks best, transfer the measurements to a rectangular piece of plywood, ½ or ¾ inch. Be sure the holes on the jig are consistent with the actual door pull. You might want to test the placement before drilling any doors. If so, you can quickly test it by mounting the pull to the jig once you have gotten the holes drilled. At this point, the jig should look like the one shown in **Figure 12-37**.

Next, nail the *stops* onto the plywood jig; this is what will hold the door pull square and properly positioned on the door (**Figure 12-38**). A top view of

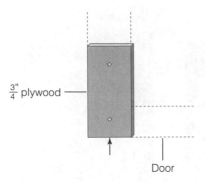

$\frac{3}{4}$" plywood

Door

FIGURE 12-37 A jig made for drilling door pulls

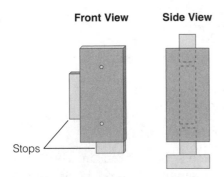

Front View   Side View

Stops

FIGURE 12-38 Stops hold the jig in position.

the jig shows where to nail on the stops; this is so that we can use the same jig for either left- or right-hand doors. Centering the stops on the plywood creates a reversible jig, one that can be used on left- *or* right-hand cabinet doors (**Figure 12-39**). When drilling for the pulls, use a bit that is slightly bigger than the screws for the pulls.

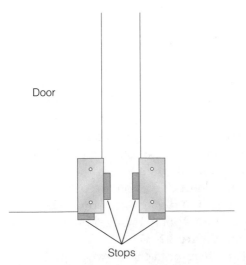

Door

Stops

FIGURE 12-39 A reversible jig can be used on right- or left-hand doors.

NOTE

When drilling through the doors, use a backing block. This will prevent the wood from splintering when the drill bit exits through the back of the cabinet door (**Figure 12-40**). ∎

Side View

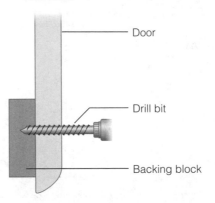

Door

Drill bit

Backing block

FIGURE 12-40 Use a backing block when drilling through doors.

CAUTION

Properly adjust blade depths (usually ⅛ inch to ¼ inch more than the thickness of the stock that is being cut).

12-5 CLOSING THE GAPS

Once the wall cabinets and the base cabinets have been installed, the **toe kicks** need to be cut and attached. Any other gaps need to be closed off as well. Normally there will be slight gaps where cabinets meet walls. These areas are covered with a screen mold or a similar molding, one that is small enough so as not to interfere with the cabinet design (**Figure 12-41**). These trims can be nailed on with small finish nails or pin nails.

Crown normally runs across the tops of the wall cabinets (**Figure 12-42**). Measure down an equal distance from the top so that the crown runs level and uniform with the tops of the doors.

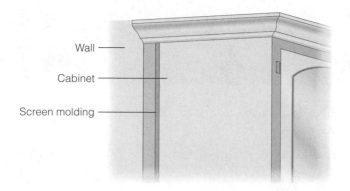

FIGURE 12-41 Screen molding that closes the gap between cabinet and wall

Wall

Cabinet

Screen molding

Gap between cabinet and ceiling

(a)

(b)

FIGURE 12-42 Crown conceals gap between top of cabinets and ceiling.

SUMMARY

- Upper cabinets are usually installed first.
- Find the highest section of floor, measure up to cabinet height, and then make a level reference line where cabinets are to be placed.
- A support board flushed with a level reference line can be used when installing upper cabinets.
- When you are installing upper cabinets, it is important for screws to hit stud centers.
- Upper cabinets are usually installed 18 inches above the finished top of the base cabinets.
- The finished height of base cabinets, including the top, is normally 36 inches.
- Installing cabinets with doors and drawers removed makes for easier installation.

- When plastic laminate is to be used on a countertop, a 1½-inch strip of wood must be added.
- Using a template when cutting the countertop for the sink ensures that the sink will be properly positioned.
- Use templates and jigs when attaching cabinet hardware.
- Use a backing board when drilling for the door pulls to prevent splintering.
- Some door patterns are designed in such a way that the door has a top and bottom; pay special attention to this detail when mounting doors and door hardware.

KEY TERMS

lowers The lower cabinets; also known as base cabinets.

scribe A mark made to indicate where material needs to be cut.

screen molding Molding usually used to cover the edges of plywood (applied to shelving edges); used to close gaps between the cabinet and wall.

toe kicks Bottom portion of the cabinet, recessed beyond the face frame.

uppers The upper cabinets, the ones mounted near the ceiling.

REVIEW QUESTIONS

1. Before installing base cabinets, the highest part of the floor must be found. How would you go about doing this?

2. What is the finished height of base cabinets?

3. Should cabinets be installed with or without the doors and drawers in place?

4. What is the typical spacing between the base cabinets and the upper cabinets?

5. What is something you can do to make setting upper cabinets easier?

6. How can you be sure that all door pulls will be attached at the same location on all the doors?

7. Should the sink be centered beneath the window (if there is one), as well as centered over the doors of the base cabinet below?

8. When drilling doors for pulls, what should you be careful of?

9. Once the upper and lower cabinets have been set, there are usually gaps in certain areas. What can be done to close the gaps where cabinets meet walls and between the tops of the cabinets and the ceiling?

10. What can be used to lift wall cabinets up and hold them in place while they are being mounted?

11. What methods can be used for marking walls at the desired height for cabinets?

12. What can be made and used as an aid for drilling door pulls?

13. What is the finished height of a vanity?

14. Ideally, when mounting cabinets, especially wall cabinets, you want to use a screw of at least what length?

Wood Flooring

CHAPTER 13

OBJECTIVES

After studying this chapter, you should be able to

- Prepare concrete slabs before installing flooring
- Install vapor barriers and sleepers over concrete slabs
- Cut and install subfloors and wood flooring
- Install transitional pieces between floors of different levels
- Install plugged flooring

INTRODUCTION

This chapter discusses the installation of hardwood and similar types of flooring (**Figure 13-1**). Today's market offers many types of wood flooring products, each of which has its own method of installation. Since carpenters traditionally work with wood and similar products, hardwood flooring primarily is discussed. Of all the types of flooring, hardwood is considered the most difficult to install. It requires more preliminary preparation, and the installation steps require a certain amount of woodworking skill. The following sections provide the fundamental information needed to complete hardwood as well as similar types of flooring. ■

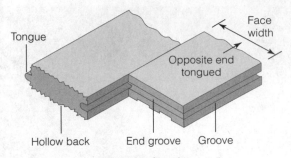

Unfinished Flooring
Sanded smooth after installation

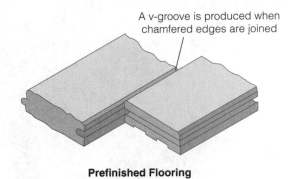

Prefinished Flooring
Chamfered edges are necessary
to obscure unevenness of surface

FIGURE 13-1 Hardwood flooring

TYPES OF HARDWOOD FLOORING

The products available for flooring range from traditional hardwood to newer products such as manufactured wood flooring. Manufactured flooring is designed with easier installation in mind for those who want wood floors but are not set up for the task of installing genuine hardwood. Other products used for flooring include woodblock and parquet, laminate, and so on.

Of all the types of flooring listed here, solid wood flooring requires the highest level of skill in terms of installation. The type of existing floor on which the hardwood is being installed affects the method of installation. Installing flooring on top of a concrete slab, for example, requires more preliminary steps than does installing flooring over a wooden **subfloor**.

MATERIALS

Hardwood flooring comes with a tongue-and-grooved edge. Flooring is available in many different species of wood. Some of the species used for hardwood flooring are ash, birch, cherry, hickory, maple, pecan, red oak, walnut, and white oak. Softwood flooring, which is installed in the same manner as hardwood flooring, is available in wood such as Douglas fir, yellow pine, and white pine.

All these types of flooring are available in a variety of widths and lengths. Widths range from 1½ to 8 inches. Lengths vary greatly. Typically, bundles of flooring are delivered with a mixture of random lengths, which helps in offsetting joints. Solid wood flooring is milled from ¼ to ¾ inch in thickness.

Manufactured wood flooring and *laminate flooring* are flooring products that are similar to plywood in construction. Laminate flooring consists of several layers of wood that have been laminated together. Manufactured wood is a tongue-and-groove material also, but since this material is dimensionally thinner than hardwood flooring, it is not typically installed with the blind-nailing method. It is installed with adhesive or allowed to float on a foam underlay. This type of flooring comes in thicknesses from ¼ to ⁹⁄₁₆ inch, in widths from 2 to 7 inches, and in lengths from 12 to 60 inches. A similar type of flooring, called **click flooring**, simply snaps together and does not require glue. It is a floating floor that rests on a foam underlay. It can be installed over existing floors (e.g. linoleum, tile, etc.).

Most of the flooring available can be purchased either prefinished or with an unfinished appearance.

Ceramic tile is a popular flooring. Most of the time, however, ceramic tile is installed by a contractor who specializes in ceramic tile. Tile work is a specialty trade requiring tools specifically designed for ceramic tile installation.

PROCEDURES: INSTALL FLOORING

SAFETY TIPS

- Use only GFCI receptacles.
- Use proper gauge extension cords.
- Never work around wet areas, and do not let extension cords run across or through wet areas.

13-1 PREPARING THE FLOOR

When putting hardwood on a concrete slab, begin by testing the slab for moisture content. A new slab will be too damp for wood flooring. At 60 days old, a slab can be tested for moisture content. Moisture testing should be performed no matter how old a slab is. Perform the test in several different rooms and areas where the flooring will be installed.

There are several ways to test a concrete slab for moisture content. One way is to lay a flat rubber mat on the floor and leave it sitting there overnight or for 24 hours. If water marks are present when the mat is removed, the slab contains too much moisture. Another way to test for moisture is to tape a 1-foot square sheet of plastic to the floor and wait 24 hours. If the plastic has become cloudy or if drops of water are present at the end of that time, the slab has too much moisture. Slabs can be chemically tested for moisture as well.

One way to speed up the drying process when slabs are too wet is with heated ventilation. Ideally, the heating and cooling system should be installed and in working order so that the structure can be climate-controlled, which is especially important once the flooring materials have been delivered. The jobsite should be climate-controlled for a week prior to delivery of flooring materials and for at least a week afterward. Perform a moisture test again once the structure has had an opportunity to dry properly.

Keep the following in mind when preparing to install hardwood flooring:

- Perform a moisture test on the slab before installing flooring.

- Generally, slabs less than 60 days old contain too much moisture for hardwood floor installation.

- Heated ventilation helps speed up the drying process for slabs with too much moisture.

- Ideally, the heating and cooling system should be installed and in working condition when you are preparing to install flooring.

Determining the moisture content of the concrete slab before installation is very important, as installing a hardwood floor over a slab that is too moist may result in damaged flooring.

13-2 PREPARING THE SLAB

The slab over which wood flooring will be placed needs to be smooth, flat, and free from oily deposits and dust. Use a grinder to take down any high spots that may be present. Fill any low spots in the floor with a leveling compound suitable for concrete foundations. Afterward, sweep the floor clean.

13-3 INSTALLING A VAPOR BARRIER OVER SLAB

It is normal for a concrete slab to maintain a certain level of moisture. However, it is important that this moisture does not reach the wood flooring. Several different systems can be used to keep moisture from reaching the finished floor.

One method is to prime the floor with an asphalt primer and then, with a **notched trowel**, apply asphalt mastic over the entire floor. Let the mastic set for a couple of hours; then, roll out 15-pound felt or rosin paper over the mastic. Lap the edges of the felt 4 inches (to the first white line). Once the entire area has been covered, apply a second coat of asphalt mastic, followed by another layer of felt building paper. Both layers of felt should be run in the same direction. Just make sure to stagger the previous overlaps, since having overlaps on top of one another will produce humps or high areas.

Another method you can use to prevent moisture from reaching a finished floor is to prime the floor

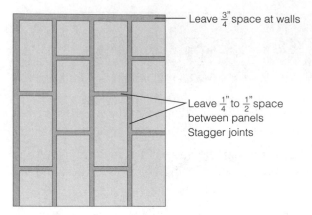

FIGURE 13-2   Stagger plywood joints and leave ¼ to ½ inch of space between panels, and a ¾ inch space at walls.

with asphalt primer, and then trowel on asphalt mastic with a **straight-edged trowel**. After an hour, lay out a 4- to 6-mil polyethylene film over the slab, lapping the edges 4 inches. Use a roller or walk over the entire surface to ensure adhesion, making sure most of the air pockets and bubbles are eliminated. Small air pockets and bubbles are harmless.

Next, use an exterior-grade plywood, at least ¾ inch thick, over the vapor barrier. Stagger the joints of the plywood, and leave ¼ to ½ inch between the plywood panels; at the walls, leave a space of ¾ inch (**Figure 13-2**). Lay the plywood at a diagonal in relation to the direction of the planned floor. Fasten the plywood to the concrete with a minimum of nine fasteners, and preferably more. (Tapcon screws are ideal for fastening plywood to the floor.) Make sure to countersink the heads of the fasteners so that they do not interfere with the flooring. Make sure the plywood is flat, start fastening in the center, and work out from there.

⚠ CAUTION

Keep work area clean, and keep blades on equipment sharp. Messy work areas and dull blades are safety hazards.

📝 NOTE

If the slab is a heated slab, you can follow the same procedure, except do not nail down the plywood. Moisture can be eliminated by turning on the slab's heating unit a week before delivery of the flooring material. ■

13-4 INSTALLING SLEEPERS OVER SLAB

Another system of preventing moisture in a slab from damaging hardwood flooring is to install **sleepers** over the slab. This method uses treated 2-inch by 4-inch boards of random lengths, between 18 and 48 inches, which are referred to as sleepers. These boards must be dry before installation.

First, clean the floor and prime it with an asphalt primer, allowing sufficient drying time. Use either hot or cold asphalt mastic, and apply it over the entire floor. Lay the sleepers perpendicular to the direction of the finished floor, 12 inches on center. Lap the ends of the boards at least 4 inches, and leave about ½ inch between where edges are lapped. This lets air circulate throughout the entire area, which helps to combat moisture. Leave a ¾-inch space at the walls. Make sure sleepers run continuously at end walls, but still leave a ½-inch gap between the boards.

Next, lay 4- to 6-mil polyethylene over the sleepers, lapping the edges over sleepers as shown in **Figure 13-3**. This system is fine for flooring that is less than 4 inches wide. Flooring that is 4 inches and wider needs to have a solid subfloor, as discussed earlier. Plywood can be laid on top of the sleepers, giving the solidity needed for flooring that is 4 inches and wider.

📝 NOTE

Insulation can be placed between sleepers. This helps to maintain a consistent temperature and cuts down on noise (e.g., the hollow sound associated with hardwood floors). ■

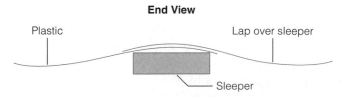

FIGURE 13-3   Polyethylene lapping over sleepers

13-5 OVER JOIST CONSTRUCTION

Considering that there is an adequate subfloor already in place, lay down 15-pound felt or rosin paper, lapping the edges 4 inches. The finished flooring can be nailed over this. If the floor is uneven, ¾-inch plywood may have to be put down; then felt or rosin paper is placed over that. Once again, the temperature needs to be maintained at near living conditions for a week prior to delivery of flooring materials so that the majority of the moisture has been eliminated before flooring installation begins.

Once most of the moisture is gone, the materials can be delivered. The flooring material will need to set for a week so that it becomes acclimated to the surrounding conditions. It helps to separate the bundle of flooring, making small bundles throughout the rooms where it is going to be installed. (Separating the material allows for better acclimation.) Once the flooring has been delivered, it is important that the temperature and humidity levels remain constant from that point on. Shutting down the cooling and heating system at any time after the flooring is on the jobsite can have a negative effect on the materials, possibly causing the flooring to shrink or buckle, depending on the conditions.

13-6 INSTALLING FLOORING

Once the floor has been prepared and vapor barriers and subfloors are in place, half the battle is over. After the flooring material has been sitting for a week and has had a chance to acclimate to the surroundings of the jobsite, you can begin installing it.

Flooring is usually laid at a right angle in relation to the floor joists. If this is the case, nails in the floor will show which direction the floor joists are running. If you are laying flooring over a large area, the starter course usually begins at the center of the room, or near the center, such as in a hallway (see **Figure 13-4**). This way the flooring can be run in both directions. Measure out an equal distance at each end of the wall, and then snap a line. Start the first run of flooring on this line, face-nailing the piece as shown in **Figure 13-5**. The **tongue** should be pointing toward the center of the room. Arrange the next six or seven runs, making sure to randomly mix longer and shorter pieces. You do not want joints to be too close together (no closer than 6 inches). Leave ½ inch of space at end walls and parallel walls. Molding will be installed later to conceal the gaps between the flooring and the walls.

It is easier to install flooring if you work in a left-to-right direction (**Figure 13-6**). If cutting a board to

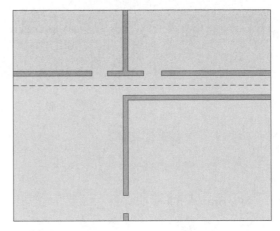

FIGURE 13-4 Begin laying flooring in the center of the house, preferably in a hallway.

FIGURE 13-5 Face-nail and blind-nail

Top View of Room

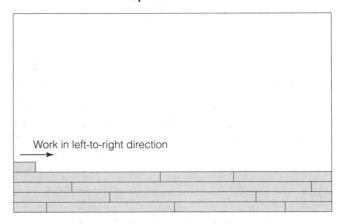

FIGURE 13-6 Install flooring in a left-to-right direction.

complete a run at the right wall, choose a board that will give you at least a 7- or 8-inch piece left over. Use the leftover piece to start the new run at the left wall. This helps to ensure a random pattern.

Make sure boards are tight against one another before nailing them. Use a scrap piece of flooring to tap in tight pieces so that the tongue is not damaged. Or you can drive a chisel into the subfloor, wedge the flooring over until snug, and then proceed to nail, using either a pneumatic or a manually operated flooring nailer (**Figure 13-7**).

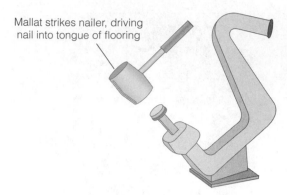

Mallat strikes nailer, driving nail into tongue of flooring

FIGURE 13-7   Flooring nailer

> ### 📝 NOTE
>
> There are manual flooring nailers as well as air-driven ones. Both produce adequate results. The manual-type nailer requires you to use a mallet, which is used to hit the head of the driving mechanism that in turn drives the nail through the flooring at the proper angle. Follow the manufacturer's recommendations as to which size and type of nail to use. Using nails of an improper size may result in a finished floor that squeaks. ■

Starter courses usually need to be face-nailed, as does flooring located in spots where a blind-nailer cannot be used (e.g., when approaching a wall). Square-edged flooring must also be face-nailed or plugged. (See the section that follows on plugged flooring.)

Flooring should run continuously through doorways and openings as shown in **Figure 13-8**. Do not permit joints at doorways. When reversing the direction of flooring, glue and insert a **spline** so that two grooves may be joined properly (**Figure 13-9**). Keep flooring ½ inch from walls. When approaching a wall, take measurements at flooring ends to make sure everything is still running squarely (**Figure 13-10**). If problems are caught in time, slight adjustments can be made so that such problems are gradually corrected.

Cut out flooring for vents and other obstacles. Flooring vents usually have flanges to cover the spots where the pieces have been crosscut across the grains. When a cut has to be made into a board in order to fit it around an obstruction, use a jigsaw and do not overcut. (See **Figure 13-11** for the steps of cutting out flooring for obstacles.)

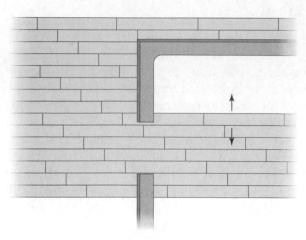

FIGURE 13-8   Flooring should run continuously through doorways.

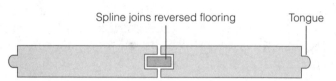

Spline joins reversed flooring   Tongue

FIGURE 13-9   A spline joins flooring when reversing direction.

A border with mitered corners can be placed around the base of the walls, fireplace hearths, and stairwells, as well as any other built-in obstacle (e.g., a bookcase), to provide framing (**Figure 13-12**).

Where floors of different heights meet, a reducer strip is needed (**Figure 13-13**). This will make the change of height between the floors less abrupt.

When nailing shoe molding around the baseboards, nail the molding into the baseboards, not into the flooring. This enables the flooring to expand and contract freely.

Measure to make sure flooring is still running square

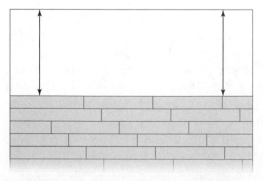

FIGURE 13-10   When approaching a wall, measure to make sure flooring is still running squarely.

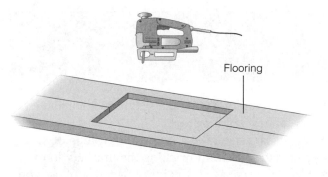

FIGURE 13-11 Use a jigsaw to cut out for vents and other obstructions, remembering not to make overcuts.

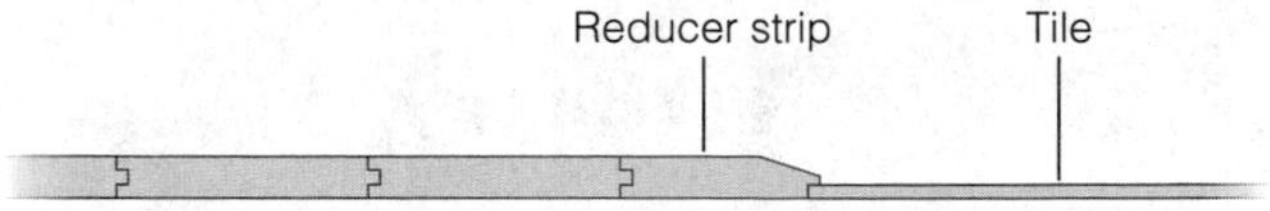

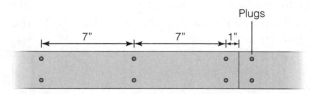

FIGURE 13-13 A reducer strip

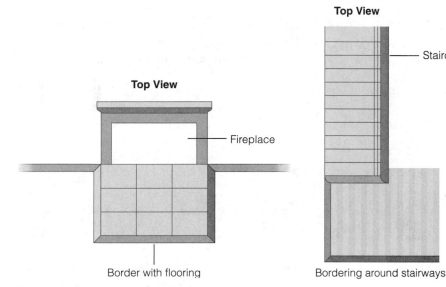

FIGURE 13-12 Flooring can be used to border fireplaces and stairwells as well as other built-in obstacles.

PLUGGED FLOORING

The look of **plugged flooring** appeals to some people. The rule of thumb when installing plugged flooring is that a 3-inch board should get one plug, a 5-inch board should get two, and a 7-inch board should receive three plugs. Preplugged flooring can be purchased or it can be made on the jobsite. Either way, you will need a counterboring bit of the size needed for the plugs you will be using.

The flooring is plugged at joints, about an inch from the ends, and is spaced about every 7 inches (**Figure 13-14**). Once the screws are in place, the plugs can be tapped in. Then, using a chisel, cut off the excess and sand plugs level with the floor (sanding with the grain of the flooring).

When installing plugged flooring, pay attention to the grain pattern of the plugs in relation to the floor's grain pattern. Decide whether the plug's grain pattern should run perpendicular (opposite) to the floor's pattern, or whether it should run parallel to (along with) the existing grain flow. Some carpenters consciously choose to set the plugs with the grains running randomly.

FIGURE 13-14 Plugged flooring, showing spacing of the plugs

SUMMARY

- Of all types of flooring offered on the market, solid wood flooring requires the highest level of skill for installation.
- Concrete slabs must be moisture-tested before putting in wood flooring.
- A new slab is usually still too wet for beginning the installation of hardwood flooring.
- Use the heating and cooling system to speed up the drying process.
- The jobsite should be climate-controlled for a week prior to the delivery of flooring materials and for at least a week after delivery of the materials.
- Slabs where flooring is going should be smooth and clean.

- It is normal for a concrete slab to maintain a certain level of moisture.
- Vapor barriers are used to keep moisture in concrete slabs from reaching the finished flooring.
- Sleepers serve as nailers for plywood subfloors on a slab.
- Insulation placed between sleepers can help maintain temperature as well as reduce hollow sounds associated with hardwood flooring.
- Flooring is usually laid at a right angle in relation to the direction of the floor joists.
- Starter courses are normally face-nailed.

KEY TERMS

click flooring A floating floor that does not require glue; it simply snaps together.

notched trowel A trowel with a notch that is used for spreading flooring adhesive or mastic. There are several different types: Some have shallow notches while other have deeper notches (⅛-inch to ⅜ inch), and some have square notches while others have triangular notches.

plugged flooring A type of flooring that has visible wooden plugs. The plugs are intended to enhance the floor's appearance. The plugs are spaced evenly, and they serve a practical purpose by covering the heads of the screws that are used to fasten down the flooring. After the plugs are glued and installed, they should be cut or sanded flush with the surface of the flooring. A Japanese Dozuki saw works well for cutting the plugs flush with the floor's surface.

sleepers Boards used to provide a nailing surface for wood flooring; they act as floor joists, keeping the wood flooring up off the concrete.

spline A thin strip of wood that is inserted into the groove of tongue-and-groove flooring when the direction of the flooring is being reversed (when the grooved ends are put together).

straight-edged trowel A trowel with a straight edge, used for spreading glue or mastic. Straight-edged trowels are not notched so that the entire surface can be completely and thoroughly covered with the material being applied. A straight-edged trowel is good to use when installing a vapor barrier over a slab.

subfloor A floor beneath the finished floor; provides a greater amount of stability.

tongue The tongue portion of tongue-and-groove material. The tongue fits snuggly into the groove.

REVIEW QUESTIONS

1. Before flooring can be installed over a slab, the slab must first be tested for what?

2. How old should a slab be before it is tested?

3. What can be done to speed up the drying process of a slab?

4. Is it normal for a slab to maintain a certain level of moisture?

5. What is put in place to stop normal slab moisture from reaching the finished flooring?

6. How long after the delivery of the flooring materials should you wait before beginning the installation process?

7. Why should materials be allowed to acclimate to the jobsite conditions?

8. What is installed between sleepers to keep the temperature constant and to reduce noise?

9. Does finished flooring typically run parallel or perpendicular to the direction of existing floor joists?

10. When two different levels of flooring meet, what can be installed to make the transition less noticeable?

11. When reversing the direction of flooring, what is glued and inserted into the grooves of the joining pieces?

12. Starter courses of flooring usually need to be nailed with what type of nailing?

13. What can be used to nail flooring?

14. Is it okay to have a joint at a doorway?

15. Explain how flooring is handled around fireplaces, stairwells, and other built-in obstacles.

16. When installing plugged flooring, how many plugs should a board that is 7 inches wide receive?

17. What is usually done between the finished floor and the baseboards to close gaps and enhance the appearance of a wood floor?

Index